T0396802

Applied Environmental Science and Engineering for a Sustainable Future

Series Editors

Applied Environmental Science and Engineering for a Sustainable Future (AESE) series covers a variety of environmental issues and how they could be solved through innovations in science and engineering. Our societies thrive on the advancements in science and technology which pave way for better and better standard of living. The adverse effects of such improvements are the deterioration of the environment. Thus, better catchment management in order to sustainably manage all types of resources (including water, minerals and others) becomes paramount important. Water and wastewater treatment and reuse, solid and hazardous waste management, industrial waste minimisation, soil and agriculture as well as myriad of other topics needs better understanding and application. The book series will aim at fulfilling such a task in coming years.

More information about this series at http://www.springer.com/series/13085

Zainul Akmar Zakaria
Editor

Sustainable Technologies for the Management of Agricultural Wastes

Editor
Zainul Akmar Zakaria
Department of Bioprocess Engineering
& Institute of Bioproduct Development,
Faculty of Chemical and Energy Engineering
Universiti Teknologi Malaysia
Johor Bahru, Johor, Malaysia

Applied Environmental Science and Engineering for a Sustainable Future
ISBN 978-981-10-5061-9 ISBN 978-981-10-5062-6 (eBook)
https://doi.org/10.1007/978-981-10-5062-6

Library of Congress Control Number: 2017961302

Printed on acid-free paper

This Springer imprint is published by Springer Nature
The registered company is Springer Nature Singapore Pte Ltd.
The registered company address is: 152 Beach Road, #21-01/04 Gateway East, Singapore 189721, Singapore

Preface

Agriculture has been the main industrial sector for many countries around the world providing millions of job to the population and is the main source of income for those countries. Nevertheless, with the increasing demand for agricultural produce, huge amounts of agricultural waste are also produced. Those agricultural wastes, also loosely termed as biomass, have huge potential to be volarized into various products such as biochemicals, biofuels and biomaterials through different physical, chemical and biological approaches mainly due to the their high carbon contents.

However, without proper management, those wastes (both liquid and solid) pose a serious threat to the environment notably from their slow degradation. Current waste management approaches are effective but would normally require a huge capital investment and are labour-intensive. They also generate potentially hazardous by-products. The above concerns prompt the need to have an alternative approach which is cheaper and easier to handle and imposes minimum adverse impact to environment.

This book provides relevant and up-to-date technologies on the utilization of various types of agricultural waste, as a direct means of properly managing its abundance. The potential of using waste materials obtained from the palm oil industry, used cooking oil, maize plantations, citrus-based plants and the tea-planting sector for the production of useful and high-value materials such as pyroligneous acid and bio-oil (Chaps. 1 and 2), biochar (Chap. 3), ferulic acid (Chap. 4) and biocontrol agents (Chaps. 5, 6, 7, 8, and 9) has been discussed. In some of the chapters, proper case studies are also included to further enhance the understanding of the readers on the subject matter highlighted. It is worth noting that even though majority of the chapters included in this book revolve around laboratory-based investigations, they have been carefully deliberated by the authors to justify their commercial feasibility which is a very important component in any research and development (R&D) ventures.

This comprehensive volume is most useful to anyone involved in agricultural waste management, green chemistry and agricultural biotechnology. It is also recommended as a reference work for all agriculture and biotechnology libraries.

Johor Bahru, Malaysia
September 2017

Zainul Akmar Zakaria

Contents

Contributors

Fatimatul Zaharah Abas Institute of Bioproduct Development, Universiti Teknologi Malaysia, Johor Bahru, Johor, Malaysia

Sharifah Soplah Syed Abdullah Section on Bioengineering Technology, Malaysian Institute of Chemical and Bioengineering Technology, Universiti Kuala Lumpur, Melaka, Malaysia

Azzam Aladdin Institute of Bioproduct Development, Universiti Teknologi Malaysia, Johor Bahru, Johor, Malaysia

Farid Nasir Ani Department of Thermo-Fluids, Faculty of Mechanical Engineering, Universiti Teknologi Malaysia, Johor Bahru, Johor, Malaysia

Suzami Junaidah Ariffin Institute of Bioproduct Development, Universiti Teknologi Malaysia, UTM, Johor Bahru, Johor, Malaysia

Lianash Azman Department of Bioprocess and Polymer Engineering, Faculty of Chemical and Energy Engineering, Universiti Teknologi Malaysia, Johor Bahru, Malaysia

Melisa Bertero Instituto de Investigaciones en Catálisis y Petroquímica "Ing. José Miguel Parera" (INCAPE) (UNL – CONICET), Santa Fe, Argentina

Julián Rafael Dib Planta Piloto de Procesos Industriales Microbiológicos (PROIMI-CONICET), Tucumán, Argentina

Instituto de Microbiología, Facultad de Bioquimica, Quimica y Farmacia, Universidad Nacional de Tucumán, Tucumán, Argentina

Hesham A. El Enshasy Institute of Bioproduct Development, Universiti Teknologi Malaysia, Johor Bahru, Johor, Malaysia

Siti Marsilawati Mohamed Esivan Department of Bioprocess and Polymer Engineering, Faculty of Chemical and Energy Engineering, Universiti Teknologi Malaysia, Johor Bahru, Malaysia

Marisa Falco Instituto de Investigaciones en Catálisis y Petroquímica "Ing. José Miguel Parera" (INCAPE) (UNL – CONICET), Santa Fe, Argentina

Li Feng Engineering Research Center for Water Pollution Source Control and Eco-remediation, College of Environmental Science & Engineering, Beijing Forestry University, Beijing, People's Republic of China

Juan Rafael García Instituto de Investigaciones en Catálisis y Petroquímica "Ing. José Miguel Parera" (INCAPE) (UNL – CONICET), Santa Fe, Argentina

Siti Zulaiha Hanapi Faculty of Chemical and Energy Engineering, Institute of Bioproduct Development (IBD), Johor Bahru, Johor, Malaysia

Ani Idris Department of Bioprocess and Polymer Engineering, Faculty of Chemical and Energy Engineering, Universiti Teknologi Malaysia, Johor Bahru, Johor, Malaysia

Institute of Bioproduct Development, Universiti Teknologi Malaysia, Johor Bahru, Johor, Malaysia

Ana Sofía Isas Planta Piloto de Procesos Industriales Microbiológicos (PROIMI-CONICET), Tucumán, Argentina

Zulsyazwan Ahmad Khushairi Faculty of Chemical and Natural Resources Engineering, Universiti Malaysia Pahang, Kuantan, Pahang, Malaysia

Umma Lawan Department of Biochemistry, Yusuf Maitama Sule University, Kano, Kano State, Nigeria

Yongze Liu Engineering Research Center for Water Pollution Source Control and Eco-remediation, College of Environmental Science & Engineering, Beijing Forestry University, Beijing, People's Republic of China

Khoirun Nisa Mahmud Institute of Bioproduct Development, Universiti Teknologi Malaysia, Johor Bahru, Johor, Malaysia

Fatima U. Maigari Department of Biochemistry, Gombe State University, Gombe, Nigeria

Roslinda Abd. Malek Institute of Bioproduct Development (IBD), Universiti Teknologi Malaysia, Johor Bahru, Johor, Malaysia

Abd. Rahman Jabir Mohd. Din Innovation Centre in Agritechnology for Advanced Bioprocess (ICA), UTM Pagoh Research Center, Pagoh Education Hub, Pagoh, Johor, Malaysia

Zulaihatu Gidado Mukhtar Department of Science Laboratory Technology, School of Technology, Kano State Polytechnic, Kano State, Kano, Nigeria

Norasikin Othman Department of Chemical Engineering, Faculty of Chemical and Energy Engineering, Universiti Teknologi Malaysia, Johor Bahru, Malaysia

Muhammad Khairul Ilmi Othman Section on Bioengineering Technology, Malaysian Institute of Chemical and Bioengineering Technology, Universiti Kuala Lumpur, Melaka, Malaysia

María Florencia Perez Planta Piloto de Procesos Industriales Microbiológicos (PROIMI-CONICET), Tucumán, Argentina

Zainab Rabiu Institute of Bioproduct Development, Universiti Teknologi Malaysia, Johor Bahru, Johor, Malaysia

Department of Biochemistry, Yusuf Maitama Sule University, Kano, Kano State, Nigeria

Roslina Rashid Department of Bioprocess and Polymer Engineering, Faculty of Chemical and Energy Engineering, Universiti Teknologi Malaysia, Johor Bahru, Malaysia

Siti Hajar Mat Sarip Faculty of Chemical and Energy Engineering, Institute of Bioproduct Development (IBD), Johor Bahru, Johor, Malaysia

Mohamad Roji Sarmidi Innovation Centre in Agritechnology for Advanced Bioprocess (ICA), UTM Pagoh Research Center, Pagoh Education Hub, Pagoh, Johor, Malaysia

Faculty of Chemical and Energy Engineering, Institute of Bioproduct Development (IBD), Johor Bahru, Johor, Malaysia

Ulises Sedran Instituto de Investigaciones en Catálisis y Petroquímica "Ing. José Miguel Parera" (INCAPE) (UNL – CONICET), Santa Fe, Argentina

Raja Safazliana Raja Sulong Institute of Bioproduct Development, Universiti Teknologi Malaysia, Johor Bahru, Johor, Malaysia

Yajun Tian Engineering Research Center for Water Pollution Source Control and Eco-remediation, College of Environmental Science & Engineering, Beijing Forestry University, Beijing, People's Republic of China

Maizatulakmal Yahayu Institute of Bioproduct Development, Universiti Teknologi Malaysia, Johor Bahru, Johor, Malaysia

Hafizuddin Wan Yussof Faculty of Chemical and Natural Resources Engineering, Universiti Malaysia Pahang, Kuantan, Pahang, Malaysia

Nor Athirah Zaharudin Department of Bioprocess and Polymer Engineering, Faculty of Chemical and Energy Engineering, Universiti Teknologi Malaysia, Johor Bahru, Malaysia

Norazwina Zainol Faculty of Chemical and Natural Resources Engineering, Universiti Malaysia Pahang, Kuantan, Pahang, Malaysia

Zainul Akmar Zakaria Department of Bioprocess Engineering & Institute of Bioproduct Development, Faculty of Chemical and Energy Engineering, Universiti Teknologi Malaysia, Johor Bahru, Johor, Malaysia

Liqiu Zhang Engineering Research Center for Water Pollution Source Control and Eco-remediation, College of Environmental Science & Engineering, Beijing Forestry University, Beijing, People's Republic of China

Yuan Zhou Engineering Research Center for Water Pollution Source Control and Eco-remediation, College of Environmental Science & Engineering, Beijing Forestry University, Beijing, People's Republic of China

Seri Elyanie Zulkifli Institute of Bioproduct Development, Universiti Teknologi Malaysia, Johor Bahru, Johor, Malaysia

Chapter 1
Introduction

Fatimatul Zaharah Abas, Seri Elyanie Zulkifli, Raja Safazliana Raja Sulong, and Zainul Akmar Zakaria

Abstract Agriculture has been the main industrial sector for various countries in the world. It provides millions of jobs to the population as well as being the main source of income for these countries. Nevertheless, with the increasing demand for agricultural produce, huge amount of agricultural waste are also produced. Using different physical, chemical, or biological approaches, these agricultural wastes, also loosely termed as biomass, present a huge potential to be valorized into various products such as biochemicals, biofuels, and biomaterials, mainly due to its high carbon composition. However, without proper management, these wastes (both liquid and solid) posed serious threat to overall environmental quality notably from its slow degradation process. Current approaches are effective but labor-intensive would normally require huge capital investment and generation of potential hazardous by-products. This prompted the need to have an alternative approach which is cheaper and easier to handle and has minimum potential impact to environmental quality.

1.1 Agriculture Industry

Agriculture is generally defined as the basis for the establishment of modern human societies. It includes crop, animal husbandry, and fisheries to fulfill the food requirements of human population (Kesavan and Swaminathan 2008; Gomiero 2017). Agriculture has emerged from merely one of the means for human survival to a

F.Z. Abas (✉) • S.E. Zulkifli • R.S.R. Sulong
Institute of Bioproduct Development, Universiti Teknologi Malaysia, Johor Bahru, Johor, Malaysia
e-mail: fatma_zahra85@yahoo.com

Z.A. Zakaria
Department of Bioprocess Engineering & Institute of Bioproduct Development, Faculty of Chemical and Energy Engineering, Universiti Teknologi Malaysia, 81310 Johor Bahru, Johor, Malaysia
e-mail: zainul@ibd.utm.my

Z.A. Zakaria (ed.), *Sustainable Technologies for the Management of Agricultural Wastes*, Applied Environmental Science and Engineering for a Sustainable Future, https://doi.org/10.1007/978-981-10-5062-6_1

commercialized, specialized, and industrialized industry. Since 1950s, the production of protein and grains has significantly increased during industrialization of agricultural system. In addition, systematic agriculture has been established in most of the developed countries in order to produce cheaper grains and proteins on smaller land area (Chen and Zhang 2015). Sharpley (2002) reported that more than 80% of agricultural product has been produced by those countries that come from 10% of farms. The relationship between agriculture and industry is vital to the overall development process and becomes a matter of interest especially in the context of recent spurt in industrial growth which has a significant impact on the economic growth and health. In most developing countries, agricultural sector has significantly contributed to the industrial and economic growth from the huge job employment opportunities (Saikia 2011). In addition, increased productivity in agricultural practice would also indirectly affect the overall economic scenario of a country where advancement in technology, knowledge in workers, agricultural practices, and management would promote the shifts in labor to higher productivity sectors which in turn offered the higher real income to them (McArthur and McCord 2017).

Agricultural waste is also known as biomass feedstock which normally has a large potential as a sustainable source of renewable energy such as for biofuel production (Demirbas 2008). Biomass energy technologies appear to be an attractive feedstock for three main reasons. First, it is a renewable resource that could be sustainably developed in the future. Second, it appears to have formidably positive environmental properties resulting in no net releases of carbon dioxide and very low sulfur content. Third, it appears to have significant economic potential provided that fossil fuel prices increase in the future (Cadenas and Cabezudo 1998). Biomass can be converted into liquid and gaseous fuels through thermochemical and biological routes. The management of agricultural waste plays an important role to achieve environmental conservation through careful planning of the waste disposal strategies. Organic wastes such as cattle manure is normally being used as fertilizers as it has been known to improve physical, chemical, and biological properties of soils. Other than that, the zeolite mineral is used as it can help in increasing the biomass production (Milosevic and Milosevic 2009). The positive effects of zeolite on nutrient status of plant and soils appear when it is mainly combined with compost or reduced doses of fertilizer.

1.1.1 Effect of Agricultural Activity

Agriculture is both critical for human well-being and a major driver of environmental decline. In fact, it is the key to attain the United Nation Sustainable Development Goals of eliminating hunger and securing food for a growing world population of 9–10 billion by 2050, which would result in the increase in global food demand between 60% and 110% (Rockström et al. 2017). The agricultural productivity is highly dependent on various factors such as soil health, natural gases (CO_2, N_2, O_2), water irrigation, and pollination insects (Kesavan and Swaminathan 2008). However,

Table 1.1 Impact of agriculture on the biophysical environment and processes (Bennett et al. 2014)

Biophysical factor	Agriculture impact
Radiative forcing	Agriculture has a significance effect on the greenhouse gas emission than any other human activity
Landfill use	50 % of savannas, grassland, and shrublands and 20 % of forest have been converted but still high pressure
Biodiversity loss	Land cover change for agriculture is one of the key drivers of biodiversity loss and could increase current extinction rates 100-fold over the twenty-first century
Freshwater	Agriculture is the largest consumer of freshwater in which almost 54 % of geographically and temporarily accessible runoff generated by Earth's hydrologic cycle each year has been consumed
Nutrients	Agriculture has greatly increased the global nitrogen and phosphorus cycles that leads to a few consequences such as human health problem, air pollution, and anoxic dead zone in freshwater and marine ecosystems

without proper management, intensive agricultural practice would lead to long-term deterioration of environmental quality notably from large emission of greenhouse gases (natural decomposition of animal's manure and excess field application of fertilizer), huge land clearing activities that would normally include burning of biomass, as well as land and water pollution, i.e., uncontrolled application of nonbiodegradable chemicals as pesticides (Kesavan and Swaminathan 2008; Bennett et al. 2014; Gomiero 2017). The impact of agriculture industries on the biophysical environment and processes is depicted as in Table 1.1.

1.1.2 Sustainable Agricultural Practice

Sustainable agricultural practice is important to reduce the impact of agricultural activities on human health, resources, and environment (Bennett et al. 2014). Indeed, the term sustainability, that basically refers to agricultural and industrial technologies that reduced or prevented environmental degradation, is often associated with economic activity. The agriculture industry ought to become increasingly sustainable at the same time as meeting society goals of access to safe, nutritious, and sufficient foods (Giovannucci et al. 2012). Among the main focus of sustainable agricultural system includes to increase the efficiency of biological process as well as agroecosystems through the development of the system-based farming that is integrated with the livestock, nutrient, water, and crop management. The ecosystem-based strategies could also be integrated with practical farm practices, where natural resources including soil, biodiversity, nutrients, and water are used as a tool to develop the productive and resilient farming systems. In addition, mixing of organic and inorganic nutrients could also be one of the approaches to manage the natural resources which are important for nutrient recycling, biological nitrogen fixation, and predation (Giovannucci et al. 2012; Rockström et al. 2017).

1.2 Agricultural Waste

Agricultural wastes can be categorized into four main groups, namely, animal waste, crop waste, food processing waste, and hazardous and toxic wastes. Agricultural wastes can be in the form of solid, liquid, or slurries depending on the raw material of agricultural products. The example of agro-waste products which includes manure, animal carcasses, cornstalks, sugarcane bagasse, drops and culls from fruits and vegetables generally has been produced based on the nature of agricultural activities (Obi et al. 2016). Furthermore, agricultural industry residues and wastes constitute a significant proportion of worldwide agricultural productivity. Even though the quantity of wastes produced by the agricultural sector is significantly lower compared to wastes generated by other industries, its deteriorative impacts must not be underestimated especially on long-term basis. For example, the impact of solid waste and hazardous waste produced from agricultural industry such as sludge from water treatment plants, chemical in fertiliser and pesticide, and used oil from agricultural machinery can cause soil contamination, air pollution through burning, odor in landfills, and water contamination (Obi et al. 2016). The proper methods of waste management need to take into account to prevent an environmental degradation, caused by inadequate disposal of these wastes. In Asia, the agricultural wastes produced are affected by socioeconomic development, land availability, and yield in crop production. The generation of agricultural waste in the Southeast Asian region is summarized as in Table 1.2 (Hsing et al. 2004).

1.2.1 Agricultural Waste Type

The types of waste produced from agricultural activities include that of cultivation activities as well as waste from livestock production. In a properly managed farm, it is estimated that approximately 30% of the feed used will eventually end up as solid

Table 1.2 Agricultural waste generation in Southeast Asia (Hsing et al. 2004)

Country	Agricultural waste generation (kg/cap/day)	Projected agricultural waste generation in 2025 (kg/cap/day)
Brunei	0.099	0.143
Cambodia	0.078	0.165
Indonesia	0.114	0.150
Laos	0.083	0.135
Malaysia	0.122	0.210
Myanmar	0.068	0.128
Philippines	0.078	0.120
Singapore	0.165	0.165
Thailand	0.096	0.225
Vietnam	0.092	0.150

waste. Manure and organic materials in the slaughterhouse, animal urine, cage wash water, and wastewater from animal slaughterhouses are a persistent source to air, water, and odor pollution (Obi et al. 2016). Other than that, the plantation of major agricultural crops such as rubber (39.67%), oil palm (34.56%), cocoa (6.75%), rice (12.68%), and coconut (6.34%) has generated huge amount of agricultural residues. From this amount, only 27.0% is used either as fuel, and the rest has to be disposed of through open burning (Salman 2015). However, the residual waste that was not utilized as fuel or did not undergo burning process, would further lead to an increase in the number of insects and weeds that would ultimately require management using available techniques such as the use of chemical-based pesticides or fungicides. Most of the packaging materials for these pesticides (bottles, canisters) would be left at the fields or ponds that can cause unpredictable environmental consequences such as food poisoning, unsafe food hygiene, and contaminated farmland due to their potentially lasting and toxic chemicals (Obi et al. 2016).

1.3 Agricultural Waste Management Strategies

Waste management is defined as the collection, transportation, processing, treatment, recycling or disposal, and monitoring of waste materials, and the term usually relates to materials produced by human activity and is generally undertaken to reduce their effect on health, the environment, or aesthetics (Demirbas 2011). Solid waste is known as the third pollution after air and water pollution, and it consists of highly heterogeneous mass of discarded materials from the urban community as well as the more homogenous accumulation of agricultural, industrial, and mining wastes. Organic wastes such as coconut shells, POME, EFB, and others that has been produced by agro-based industry are not hazardous in nature and thus have potential for the other uses. These agricultural wastes can be converted to useful products such as biochemicals, biofuels, and biomaterials using different approaches (Tahir 2012; Abdullah and Sulaiman 2013). The process of selection for the right solid waste disposal method is complex due to high heterogeneous mass involved (Table 1.3; Gertsakis and Lewis 2003).

The conventional waste management hierarchy is focused on disposal via landfilling and/or incineration with a minimum effort of extracting value by recycling

Table 1.3 Waste management hierarchy (Gertsakis and Lewis 2003)

Goal	Attribute of ways	Outcomes
Avoid and reduce	Preventative	Most desirable
Reuse, recycle, recover energy	Predominantly ameliorative, part of preventative	↕
Treatment	Predominantly assimilative, partially ameliorative	
Disposal	Assimilative	Least desirable

and/or energy recovery from waste. This hierarchy consists of options for waste management during the lifecycle of waste, arranged in descending order of priority (Pariatamby and Periaiah 2010). There are six functional elements of waste management in conventional technique, namely, generation, handling, collection, transfer, recovery, and disposal. Out of these, the three basic functional elements of solid waste management are collection, treatment, and disposal.

References

Abdullah N, Sulaiman F (2013) The oil palm wastes in Malaysia. Biomass Now-Sustain Growth Use 1(3):75–93

Bennett E, Carpenter S, Gordon LJ, Ramankutty N, Balvanera P, Campbell B, Cramer W, Foley J, Folke C, Karlberg L, Liu J (2014) Towards a more resilient agriculture. Solutions 5(5):65–75

Cadenas A, Cabezudo S (1998) Biofuels as sustainable technologies: perspectives for less developed countries. Technol Forecast Soc Chang 58(1):83–103

Chen HG, Zhang YHP (2015) New biorefineries and sustainable agriculture: increased food, biofuels, and ecosystem security. Renew Sust Energ Rev 47:117–132

Demirbas A (2008) Biofuels sources, biofuel policy, biofuel economy and global biofuel projections. Energy Convers Manag 49(8):2106–2116

Demirbas A (2011) Waste management, waste resource facilities and waste conversion processes. Energy Convers Manag 52(2):1280–1287

Gertsakis J, Lewis H (2003). Sustainability and the waste management hierarchy. Retrieved on January, 30, 2008

Giovannucci D, Scherr SJ, Nierenberg D, Hebebrand C, Shapiro J, Milder J, Wheeler K (2012). Food and agriculture: the future of sustainability

Gomiero T (2017). Agriculture and degrowth: state of the art and assessment of organic and biotech-based agriculture from a degrowth perspective. J Clean Prod

Hsing HJ, Wang FK, Chiang PC, Yang WF (2004) Hazardous wastes transboundary movement management: a case study in Taiwan. Resour Conserv Recycl 40(4):329–342

Kesavan PC, Swaminathan MS (2008) Strategies and models for agricultural sustainability in developing Asian countries. Philos Trans of the R Soc of Lond B Biol Sci 363(1492):877–891

McArthur JW, McCord GC (2017) Fertilizing growth: agricultural inputs and their effects in economic development. J Dev Econ 127:133–152

Milosevic T, Milosevic N (2009) The effect of zeolite, organic and inorganic fertilizers on soil chemical properties, growth and biomass yield of apple trees. Plant Soil Environ 55(12):528–535

Obi FO, Ugwuishiwu BO, Nwakaire JN (2016) Agricultural waste concept, generation, utilization and management. Niger J Technol. 35(4):957–964

Pariatamby A, Periaiah N (2010) Waste management challenges in sustainable development of Islands. In: International Solid Waste Association (ISWA) World Congress

Rockström J, Williams J, Daily G, Noble A, Matthews N, Gordon L, Wetterstrand H, DeClerck F, Shah M, Steduto P, de Fraiture C (2017) Sustainable intensification of agriculture for human prosperity and global sustainability. Ambio 46(1):4–17

Saikia D (2011) Trends in agriculture-industry interlinkages in India: pre and post-reform scenario. In: Saikia D (ed) Indian economy after liberalisation: performance and challenges. SSDN Publication, New Delhi, pp 122–173

Sharpley AN (2002) Introduction: agriculture as a potential source of water pollution. In: Haygarth PM, Jarvis SC (eds) Agriculture, hydrology, and water quality. CABI Publishing, Cambridge, MA

Tahir TA (2012) Waste audit at coconut-based industry and vermicomposting of different types of coconut waste. Ph.D. thesis, University of Malaya

Salman Zafar (2015) Agricultural biomass in Malaysia. BioEnergy Consult

Chapter 2
Pyrolysis Products from Residues of Palm Oil Industry

Melisa Bertero, Juan Rafael García, Marisa Falco, Ulises Sedran, Khoirun Nisa Mahmud, Suzami Junaidah Ariffin, Ani Idris, and Zainul Akmar Zakaria

Abstract Palm kernel shell (PKS) and empty fruit bunches, both raw (EFB-R) and pretreated by means of autoclaving (EFB-A) and microwave (EFB-M), were pyrolyzed in a fixed-bed stainless steel reactor at 550 °C. The yield of the water-soluble liquid fraction (pyroligneous acid, PA) in the pyrolysis of PKS was 26%wt. (dry basis) and in the range of 16–46%wt. when different EFB were used. The yield of insoluble liquid fraction (bio-oil, BO) was 9.1%wt. for PKS and up to 25%wt. in the case of EFB. Liquid and gaseous products were analyzed by conventional capillary gas chromatography. The PA from the PKS had 30%wt. of total phenolic compounds (up to 24%wt. phenol) and 46%wt. acetic acid. On the other hand, the bio-oil from PKS had 43%wt. of total phenolic compounds (up to 26%wt. phenol) and

M. Bertero • J.R. García • M. Falco • U. Sedran (✉)
Instituto de Investigaciones en Catálisis y Petroquímica “Ing. José Miguel Parera” (INCAPE) (UNL – CONICET), Santa Fe, Argentina
e-mail: mbertero@fiq.unl.edu.ar; jgarcia@fiq.unl.edu.ar; mfalco@fiq.unl.edu.ar; usedran@fiq.unl.edu.ar

K.N. Mahmud • S.J. Ariffin
Institute of Bioproduct Development, Universiti Teknologi Malaysia, Johor Bahru, Johor, Malaysia
e-mail: khoirunnisa.m@gmail.com; suzamijunaidah@gmail.com

A. Idris
Instituto de Investigaciones en Catálisis y Petroquímica “Ing. José Miguel Parera” (INCAPE) (UNL – CONICET), Santa Fe, Argentina

Institute of Bioproduct Development, Universiti Teknologi Malaysia, Johor Bahru, Johor, Malaysia
e-mail: ani@cheme.utm.my

Z.A. Zakaria
Department of Bioprocess Engineering & Institute of Bioproduct Development, Faculty of Chemical and Energy Engineering, Universiti Teknologi Malaysia, 81310 Johor Bahru, Johor, Malaysia
e-mail: zainul@ibd.utm.my

Z.A. Zakaria (ed.), *Sustainable Technologies for the Management of Agricultural Wastes*, Applied Environmental Science and Engineering for a Sustainable Future, https://doi.org/10.1007/978-981-10-5062-6_2

17%wt. acetic acid. The PA from EFB contained mainly acetic acid (65.5%wt.), furfural (7.7%wt.), methanol (8.0%wt.), and phenol (15.2%wt.). When EFB was pretreated, the concentration of acetic acid in PA decreased dramatically, while the concentration of furfural increased up to ten times, this effect being more noticeable in the case of microwave pretreatment. The yields of by-products were of significance in all cases (13–23%wt. of gases and 33–52%wt. of char). These results show that the liquid products obtained from the pyrolysis of palm oil industrial wastes could be used in order to obtain chemical raw materials of worldwide extended use, while the by-products (gases and char) can be used as renewable energy sources.

2.1 Introduction

Residual lignocellulosic biomass is a renewable, low price resource which could be used for the production of energy, transportation fuels, and chemicals (Dutta et al. 2012). The pyrolysis (thermochemical degradation process with limited amount of oxygen) of this biomass allows obtaining up to 70%wt. of liquid products: pyroligneous acid (PA, water-soluble fraction) and bio-oil (BO, water-insoluble fraction). The thermal process permits producing useful chemicals from waste lignocellulosic biomass from agriculture, forestry, or industrial wastes (Bertero and Sedran 2014). Moreover, it does not compete with food production, and the raw materials are inexpensive. The PA is a very complex mixture of oxygenated compounds (more than 40% wt. oxygen) and water, while the BO contains oxygenated compounds with higher molecular weights.

Freel and Graham (2002) proposed that the liquid product from the pyrolysis of hard wood chips can replace creosote (a wood preservative obtained from coal tar), because some phenolic compounds in its composition showed to have insecticide and fungicide properties. Due to the large number of compounds with different chemical functionalities in the liquid products from pyrolysis, the isolation of different fractions is certainly a complex matter. However, many techniques were described to separate the phenol-rich fraction by precipitation or liquid-liquid extraction (Effendi et al. 2008). The insoluble fraction (BO, rich in phenolic compounds up to 30–50%wt. from the lignin) has interesting industrial uses. For instance, it can partially replace phenol in phenol-formaldehyde resins, resulting in lesser environmental impact and lower toxicity and costs. It is also possible to separate compounds with industrial interest, such as furfural, which finds numerous applications as an important renewable chemical feedstock, used as chemical solvent, extractant, fungicide, nematocide, or precursor of different chemicals (Ani 2015). Hydrogenating furfural provides furfuryl alcohol, which is a useful chemical intermediate in many instances, for example, the manufacture of furan resin prepolymers in thermoset polymer matrix composites, cements, adhesives, casting resins, and coatings. The further hydrogenation of furfuryl alcohol produces tetrahydrofurfuryl alcohol, a nonhazardous solvent in agricultural formulations and an adjuvant in herbicides. Furfural is also used to make other furan chemicals, such as furoic acid, via oxidation, and furan itself via palladium-catalyzed vapor-phase decarbonylation.

Palm oil is the edible oil with the highest consumption in the world, and Malaysia is the second largest producer, with 39% of the total palm oil world production in 2010/2011. The total area cultivated with oil palm in Malaysia is estimated in about 5.64 million hectares in 2015 (MPOB 2015). The palm oil industry generates large amount of biomass wastes such as palm kernel shell (PKS, Malaysia, 4.8 million ton/year), oil palm empty fruit bunch (EFB, Malaysia, 22.1 million ton/years), and palm oil mill effluents (POME). All these wastes could be used to produce biofuels and chemical raw materials; particularly, some reports showed that the palm oil biomass is a raw material adequate to produce various important chemicals, such as succinic acid (Akhtaret al. 2014a, b) or lactic acid (Ani 2015).

Prior to pyrolysis, various treatments can be applied to the lignocellulosic wastes in order to, e.g., reduce cellulose crystallinity, increase biomass porosity, or facilitate the separation of the major components (cellulose, hemicellulose, and lignin), thus improving the yields of certain compounds. The conditioning treatments could be either physical with steam or chemical with alkalis and/or acids (Bhatia et al. 2012). In general terms, alkalis are more efficient to separate lignin when the amount of lignin in biomass is low and acids when it is high (Mosier et al. 2005). It has been reported that alkali treatments allow removing lignin by scission of intermolecular bonding between lignin and hemicellulose, while acids degrade hemicellulose (Ani 2015). Moreover, if these treatments are microwave assisted, the effectiveness is higher, as shown in the cases of rice straw (Zhu et al. 2006), switch grass (Hu et al. 2008), and EFB (Hamzah et al. 2009).

The possible upgrading scheme of oil palm lignocellulosic wastes is shown in Fig. 2.1, and, in view of the possible use of these residuals in thermochemical processes in order to obtain chemicals and fuels, it is the purpose of this work to determine the distribution of products in the pyrolysis of different palm oil residuals, emphasis being given to the composition of the liquid products.

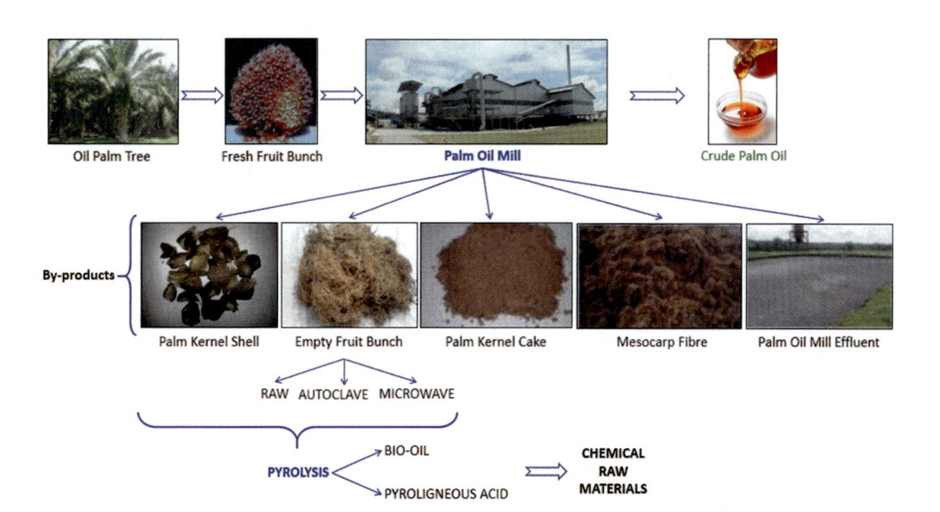

Fig. 2.1 Schematic representation of thermochemical processing of lignocellulosic wastes from oil palm industry

2.2 Pyrolysis of Palm Oil Industry Wastes

2.2.1 Materials

Four biomass raw materials were used: palm kernel shell (PKS) and three samples of empty fruit bunches (EFB). The PKS sample was obtained from a palm oil mill in Johor, Malaysia (Kulai Palm Oil Mill). It was washed with tap water to remove soil, dust, and other unwanted materials prior to sunlight drying during day 1 to remove the external moisture and prevent the growth of fungi during storage. The dried material was ground and sieved to obtain particles with approximately 2–4 mm size. Prior to pyrolysis, the PKS sample was dried again in an oven to remove the excess moisture at 105 °C for 24 h. The EFB biomass was collected from FELDA palm oil mill Semenchu, Kota Tinggi, Johor, Malaysia, washed with tap water, and subsequently sun dried. The EFB biomass was milled to approximately 0.5 × 1.0 cm size using a disk mill model FFC-15, China, and sieved using a Retsch sieve shaker (AS 200 basis, Germany) to less than 1.0 mm particle size (mesh size <N° 18). Prior to analysis, different powdered samples were dried in an oven at 60 °C for 24 h and stored in an airtight container at room temperature (Akhtar et al. 2014a, b). One of the EFB samples was used without previous treatment (EFB-R), a second sample was pretreated by means of an autoclave method (EFB-A), and the third sample was pretreated using microwave irradiation (EFB-M). In autoclaving, 20 g of dry EFB was soaked with 100 ml of 2.5 M NaOH solution during 2 h and then heated in autoclave at 121 °C during 1 h; this alkali-treated EFB (EFB-A) was then separated from the alkaline solution, washed with tap and distilled water, filtered, and dried at 90 °C overnight (Akhtar et al. 2014). In microwaving, 20 g of dry EFB was soaked in 100 ml of 8.0%v. H_2SO_4 solution and heated in autoclave at 121 °C during 1 h and then washed with distilled water and dried in an oven at 110 °C overnight. The dried EFB was then soaked in 2.5 M NaOH solution (1 g solid/10 mL solution) and then subjected to microwave radiation. A SINEO's microwave oven (MAS-II, China) was used at microwave power 700 W and 80 °C during 60 min. The microwave-treated biomass was washed with water and then dried in an oven 110 °C overnight prior to subsequent experiments. The various biomasses were characterized following standard methods: extractives with Soxhlet extraction, Klason lignin and holocellulose (TAPPI 2002; T222 om-02), and cellulose (T203 cm-09). Hemicellulose was determined by difference between holocellulose and cellulose.

2.2.2 Pyrolysis Experiments

The experiments of pyrolysis were performed in a fixed-bed reactor (stainless steel, 21 cm long, 1.9 cm diameter) in an electrically heated oven (Bertero et al. 2011). A heating ramp of 20 °C/min was used from room temperature to 550 °C, and the final temperature was maintained at a constant value during 30 min. A 30 ml/min nitrogen flow was used in order to sweep the vapors from the reaction zone and to minimize the undesirable secondary condensation and carbonization reactions. The reactor effluents passed through a condenser immersed in a saline solution at −5 °C, where liquid products (pyroligneous acid and bio-oil) were retained, and the gases were collected in a water column and quantified by displacement. After the reaction time was completed, the reactor was swept with nitrogen during 7 min. Mass balances (recoveries) were higher than 90% in all the cases.

2.2.3 *Product Analysis*

The liquid products were separated by centrifugation at 3200 rpm during 20 min in order to obtain the pyroligneous acid (PA) and bio-oil (BO) fractions. Both PA and BO were analyzed by means of conventional capillary gas chromatography in an Agilent 6890 N chromatograph with a 30 m long, 250 μm diameter, and 0.25 μm film thickness, dimethylpolysiloxane HP-5 column, and flame ionization detection (FID). In order to perform the GC analysis, the BO fraction was dissolved into dichloromethane. The identification of products was performed with the help of standards and comparisons with previous works (Bertero et al. 2012, 2014) and by means of GC-MS. In the GC-MS analysis, an Agilent system (6890 N, Agilent Technologies CA, USA) equipped with a mass selective detector (5973 N) was used in the electron ionization mode. The identification of the components was based on the comparison of their mass spectra with the NIST Search Library 2.0 as well as by comparison of their retention times with existing literature data (Mathew et al. 2015). Gases were analyzed in an Agilent 6890 N gas chromatograph with a 30 m long, 530 μm diameter, and 3.0 μm film thickness, bonded monolithic carbon-layer GS-CARBONPLOT column, with thermal conductivity detection (TCD). Peak areas were calibrated using response factors specific for each of the chemical groups, in turn determined with mixtures of standards and reference compounds (tetralin for liquids and methane for gases).

The water content in PA was determined by means of Karl-Fischer titration (IRAM 21320 standard).

The yields of the various products were calculated as the relationship between the mass of the product and the mass of dried biomass used. The composition of the carbonaceous residue remaining after the pyrolysis experiments was evaluated by means of elemental analysis by combustion in a CHN628 Series equipment (LECO), and the calorific power (higher heating value, HHV) was calculated by means of the Dulong's formula (Özbay et al. 2008).

2.3 Characterization of Raw and Modified Palm Oil Industry Wastes and Their Pyrolysis Products

2.3.1 Characterization: Comparison with Other Waste Biomasses

The lignocellulosic wastes from oil palm industry are mainly composed by cellulose, hemicellulose, and lignin, together with minor amounts of inorganic minerals, proteins, lipids, simple sugars, and starch. For example, according to Hamzah et al. (2009), EFB contains 44.2%wt. cellulose, 33.5%wt. hemicellulose, and 20.4%wt. lignin. PKS contains 33.04%wt. cellulose, 23.82%wt. hemicellulose, and 45.59%wt. lignin (Kim et al. 2010). POME, an effluent generated from palm oil processing contains high biochemical oxygen demand (BOD) (25,000–65,714 mg/L), chemical oxygen demand (COD) (44,300–102,696 mg/L), total solids (40,500–72,058 mg/L), suspended solids (18,000–46,011 mg/L), and volatile solids (34,000–49,300 mg/L) (Chin et al. 2013).

Conditioning treatments on these waste biomasses are performed aimed at disrupting the lignocellulosic structure and facilitate further processing, for example, enzymatic digestion to produce lactic or succinic acids or fermentation to produce alcohols or to maximize the yields of certain compounds by means of thermochemical methods. The treatment with NaOH saponifies the intramolecular ester bondings, which are responsible of cross-linking in hemicellulose, xylan, and lignin, thus increasing biomass porosity to ease its further pyrolysis (Sun and Cheng 2002). Contacting biomass with sulfuric acid is efficient in removing hemicellulose but not lignin (Ani 2015).

Treatments on EFB decreased considerably the content of lignin in biomass, particularly microwave irradiation, where it was possible to remove up to 72% of lignin. This treatment also removed 90% of hemicellulose in the acid treating step before microwave heating. Previous studies determined the best conditions for microwave EFB delignification (temperature, microwave power, and time) (Akhtar et al. 2014) and showed that the most important parameter is microwave power. Consequently, treated EFB, particularly EFB-M, contained mainly cellulose (Table 2.1).

2.3.2 Global Yields and Composition of Liquid and Gaseous Products in the Pyrolysis: Comparison with Other Waste Biomasses

After the implementation of the policy on greenhouse gas (GHG) emission reduction, PKS has been broadly used as a renewable fuel in Malaysia in the past 10 years (Dit 2007). However, an alternative approach is using PKS in replacement of fossil

Table 2.1 Composition of oil palm waste biomass – palm kernel shell, PKS, and empty fruit bunch raw, EFB-R (%wt., dry basis)

	PKS	EFB-R
Fixed carbon	20.3	21.7
Volatile material	77.2	72.3
Cellulose	33.04	
Hemicellulose	23.82	
Lignin	49.77	18.9
Extractives	9.89	
Ash	2.5	6.0
Elemental composition		
C	55.4	47.9
H	6.3	6.4
O	38.0	45.4
N	0.4	0.7
HHV (MJ kg $^{-1}$)	19.6	17.8

Table 2.2 Yields of pyrolysis products at 550 °C (%wt., dry basis)

	PKS	EFB-R	EFB-A	EFB-M
Bio-oil (BO)	9.1	–	24.8	21.8
Pyroligneous acid (PA)	26.0	46.2	19.2	16.8
Water content in PA	51.2	55.8	77.1	73.8
Char	51.7	35.0	32.8	41.1
Gases	13.2	18.8	23.1	20.3

resources, such as oil and natural gas, to obtain chemicals of broad use in the industry or fuels. EFB is used in compost production, but some studies showed that is possible to consider it as a carrier in the inoculation of bacteria in the formulation of biofertilizers (Hoe et al. 2016).

The yields of the different products in pyrolysis processes mainly depend on the raw material and the reaction temperature. According to various reports from other authors, there exists a maximum in the bio-oil yield as a function of temperature, typically close to 550 °C (Williams and Besler 1996; Pütün et al. 2002; Bertero et al. 2011). Moreover, the heating rate in the 5–80 °C/min range has no significant effect on the product distributions (Özbay et al. 2008).

Table 2.2 shows the yields of different products obtained in the pyrolysis experiments of the various biomass sources. The EFB-R yielded more liquid products than PKS, and the delignification treatments favored the yield of bio-oil in detriment of that of pyroligneous acid (PA). In all the cases the PA had a high content of water, but, particularly in the case of EFB, the water content in the treated samples (both by autoclave and microwave methods) was approximately 50% higher than that in the raw, untreated sample. It is clear that these treatments contribute to ease the dehydration reactions during the subsequent pyrolysis process, because the solid remaining after the treatments is richer in cellulose, which is more susceptible to dehydrate

(Stefanidis et al. 2014). The lower water content in PA from EFB-M, as compared to that derived from EFB-A, could be due to the lower content of hemicelulose in the biomass treated with microwaves since, at temperatures higher than 500 °C, xylan, which is the major primary product in hemicelluloses pyrolysis, dehydrates easily, thus increasing considerable the water yield (Stefanidis et al. 2014).

The yield of char in the pyrolysis of PKS was much higher than that in the case of the EFB, while gases were produced more significantly with EFB, specially with the pretreated samples. Both by-products, char and gases, could be used as energy sources in self-sustaining large-scale processes of pyrolysis aimed at producing liquid products, from which chemicals could be obtained.

The pyrolysis of other residual biomasses derived from oil palm exploitations, such as pericarp from the white palm fruit and the residue from the extraction of oil from white palm seeds, under the same experimental conditions, yielded between 3 and 8.8%wt. bio-oil and between 32 and 35%wt. pyroligneous acid (Bertero et al. 2014). In those cases, gases represented between 20 and 26%wt. of the raw biomass and char between 36 and 38.5%wt. Other authors reported similar values for the product streams in the conventional pyrolysis of various raw materials, such as wood sawdusts, fruit pulps, crop residue, and fruit shells (Kawser and Farid Nash 2000; Demirbas 2001; Pütün et al. 2002, 2004, 2005; Özbay et al. 2008).

The pyroligneous acids are extremely complex aqueous mixtures made up of a large number of compounds from various chemicals species derived from the depolymerization and fragmentation of the three main components in biomass: cellulose, hemicellulose, and lignin. On the contrary, bio-oil is the water-insoluble fraction in the pyrolysis liquid products; it mainly contains products from the degradation of lignin (Bayerbach and Meier 2009). In this sense, Table 2.3 shows the composition of PA and BO liquid products in the pyrolysis of different biomasses.

The main chemical compounds present in the PA from the untreated biomasses were acetic acid, phenol, methanol, and furfural. Particularly the amount of phenol in the liquid products from PKS pyrolysis (both PA and BO) was very high due to the higher content of lignin in this biomass. These evidences suggest that PA and BO derived from PKS would be potentially useful to partially replace phenol in the formulation of phenol-formaldehyde resins, for example. Some studies show that it would be possible to use these liquids as the substitutes of phenol from fossil sources, without the need for upgrading (Kawser and Farid Nash 2000).

When the EFB were pretreated, the concentration of acetic acid in PA decreased significantly while that of furfural increased up to ten times, the effect being more noticeable in the case of EFB-M. The concentration of phenol in PA was the same after the treatments, resulting about 15%wt. in all the cases, even though the decline in the content of lignin had been very important (see Table 2.1). However, as already discussed (see Table 2.1), the yield of PA decreased significantly when the EFB were pretreated, thus making phenol yield to drop up to 60%.

The content of acetic acid in PA from EFB pretreated with acid and microwave (EFB-M) reduced by 95% compared with that of EFB-R, because acetic acid derives mainly from the deacetylation of hemicellulose (Alén et al. 1996), and the treatments eliminated 90% of hemicellulose in EFB (see Table 2.1). In general, pyroligneous

Table 2.3 Composition of liquid products in the pyrolysis of PKS and treated and untreated EFB biomasses (%wt., dry basis)

	PKS		EFB		
	PA	BO	PA (R)	PA (A)	PA (M)
Acetic acid	49.4	24.0	65.5	12.6	3.5
Decanoic acid	0.1	1.2	–	–	–
Benzoic acid, 4-hydroxy-	0.2	–	–	–	–
Acetaldehyde	–	–	0.4	1.6	0.1
Acetone	0.8	0.7	1.6	2.1	0.8
2,3-Pentadione	–	0.4	0.3	1.2	0.4
2-Penten-3-one	–	0.7	0.4	0.2	0.1
Furfural	6.3	8.8	7.7	56.2	76.9
Methanol	10.7	4.6	8.0	11.5	2.5
Octyne	0.1	0.4	0.9	0.2	0.8
Phenols	**31.9**	**59.4**	**15.2**	**14.4**	**15.0**
Phenol	26.1	36.3	7.2	2.8	5.9
o-Cresol	0.5	1.4	0.5	0.8	0.9
m-Cresol	0.2	0.6	0.8	0.3	0.8
p-Cresol	0.5	1.4	0.4	0.3	0.9
Phenol, 2,4-dimethyl-	0.1	0.6	0.5	6.3	1.0
Phenol, 3 ethyl-	0.1	0.3	0.4	–	–
o-Guaiacol	1.6	4.1	0.7	–	–
4-Methylguaiacol	0.8	3.3	0.7	0.2	0.5
4-Ethylguaiacol	0.1	0.7	0.5	3.1	0.5
Syringol	0.6	2.3	0.8	0.2	0.5
Eugenol	0.1	0.6	0.5	0.1	0.4
Phenol, 2-ethoxy-	0.1	0.4	0.3	–	–
Phenol, 2-methoxy-4-propyl-	0.4	0.3	0.3	0.1	0.3
Benzoic acid, 4-methoxy-, methyl ester	0.2	0.3	0.7	–	0.6
Benzaldehyde, 3-hydroxy-4-methoxy-	0.2	2.9	0.7	–	1.2
Ethanone, 1-(3-hydroxy-4-methoxyphenyl)-	0.1	0.8	0.3	–	1.4
4-Ethyl-1,3-benzenediol	–	2.9	–	–	–
3-Methoxycatechol	0.5	–	–	–	–

acids contain high amounts of acetic acid, for example, between 15 and 59%wt. when obtained from various wood sawdusts and fruit shells (Valle et al. 2012) and between 23 and 46%wt. biomasses derived from the upgrading of white palm wastes (Bertero et al. 2014).

It has also been reported that previous alkali or acid treatments on biomasses reduce the degree of crystallinity and polymerization of cellulose (Mosier et al. 2005), a fact that eases its pyrolysis. Among products in the pyrolysis of cellulose is levoglucosan, which can undergo dehydration and isomerization reactions to form other anhydrosugars, including levoglucosenone and 1,6-anhydro-β-d-glucofuranose. The anhydrosugars can form furans, such as furfural and hydroxymethylfurfural, by

Table 2.4 Composition and properties of the gaseous products in the pyrolysis

	PKS	EFB-R	EFB-A	EFB-M
Chemical composition (%wt.)				
Methane	4.3	4.1	3.5	3.8
Ethane + ethylene	0.5	0.5	1.3	1.5
Propane + propylene	0.3	0.3	4.2	4.7
CO_2	87.7	89.2	81.1	78.5
CO	2.7	2.0	0.1	2.5
Hydrogen	3.8	3.3	4.8	3.4
Elemental composition (%wt.)				
C	29.3	29.2	31.8	33.2
H	5.0	4.5	7.0	5.8
O	65.3	66.0	59.0	58.5
HHV (MJ kg^{-1})	9.0	8.1	12.6	11.1

dehydration reactions, or hydroxyacetone, glycolaldehyde, and glyceraldehydes, by fragmentation and retro-aldol condensation reactions (Lin et al. 2009).

In the case of BO from PKS, the main compounds were essentially the same as those in the corresponding PA, but the concentration of phenol was much higher. Many phenolic compounds were also present in significant proportions in the BO, like o-guaiacol, 4-methylguaiacol, 3-hydroxy-4-methoxy-benzaldehyde, and 4-ethyl-1,3-benzenediol. Guaiacol and its alkylated homologous compounds are usually the most important phenolic ethers present in bio-oils (Egsgaard and Larsen 2000; Hosoya et al. 2007). Other bio-oils derived from the upgrading of white palm fruit wastes contain between 33 and 38%wt. of phenolic compounds, the most important ones being phenol, cresol, guaiacol, and its alkylated homologous compounds, isoeugenol and siringol (Bertero et al. 2014).

All the major products in the pyrolysis liquids, which are in general produced from oil, such as phenol and furfural, and natural gas, such as methanol, are useful chemical raw materials with a large number of industrial applications.

Gases which are produced in pyrolysis processes include carbon oxides, hydrogen, and light hydrocarbons. They can be used at the commercial level as fuels in the own pyrolysis process, providing part of the heat required for the endothermic reactions. It has been shown that hemicellulose contributes particularly to CO_2, cellulose to CO, and lignin to H_2 and CH_4, according to the different mechanisms which control their conversions (Alén et al. 1996; Collar and Blin 2014). Table 2.4 shows that gaseous products in all the pyrolysis were mainly composed of carbon dioxide, methane, and hydrogen, with PKS and EFB-R producing more methane due to their higher lignin content, as compared to pretreated EFBs.

EFB pretreatments caused CO_2 concentration to decrease and C_2-C_3 hydrocarbons and hydrogen concentrations to increase in the gaseous products from pyrolysis. Consequently, those gases contained more C and H, increasing HHV, and less O than those derived from untreated PKS and EFB biomasses.

It is known (Alén et al. 1996; Collar and Blin 2014) that the different components of biomass show different yields to solid products in pyrolysis; for example, hemicellulose produces about 7%wt. char, while this yield increases up to 45%wt. in the case of lignin. PKS produced more char than EFB-R as a consequence of its higher lignin content. For example, the composition of the carbonaceous residue after the pyrolysis of PKS included 60.9%wt. C, 34.6%wt. O, 2.6%wt. H, 0.8%wt. N, and 1.1%wt. ash., its HHV being 18.3 MJ kg^{-1}. The proportion of carbon and oxygen in the char from PKS can be compared to be lower and higher, respectively, than the values corresponding to chars derived from wood sawdusts (e.g., pine sawdust char includes 90.9%wt. C and 7.3%wt. O (Bertero et al. 2011)). Consequently, its HHV was lower than those reported for wood sawdusts (e.g., 32.0 MJ kg-1 in the case of pine sawdust (Bertero et al. 2011)). Besides its use as a fuel in the own pyrolysis, char can be used in, for example, as a toxin trapping agent in livestock food, as soil fertilizer, or could even be gasified in a further step in the reactor to obtain synthesis gas, which in turn could be feasibly used as an energy source.

2.3.3 *Potential Applications of Pyroligneous Acid and Bio-oil*

Pyroligneous acids from pyrolysis find applications in various areas as antioxidants, antimicrobial, or anti-inflammatory agents, plant growing stimulants, natural rubber coagulants, termiticide and pesticide agents, and smoky flavors for food, but they are also a source of valuable chemicals (Mathew et al. 2015). PA from the pyrolysis of lignocellulosic wastes from oil palm industry could be used as preservative in forestry industries, considering their high concentrations of acetic acid and phenolic compounds. A publication by Wititsiri et al. (2011) showed that acetic acid in PA from coconut shell pyrolysis has a strong termiticide activity, and the same effect was observed by Yatagai et al. (2002) for phenols in PA from sawdust pyrolysis from various woods. Also acetic acid and phenols are the most important active components in PA when used as food flavor additive, while phenolic compounds, which, in general, showed antioxidant activity, were shown to be useful in the formulation of dermatological products (Kimura et al. 2002) and synthetic antioxidants (Loo et al. 2007). Acetic acid could be recovered from PA by means of liquid-liquid extraction with aliphatic tertiary amines. Liquids from the pyrolysis of lignocellulosic wastes from oil palm industry, particularly those from PKS could be used as raw materials to separate some phenolic compounds, such as guaiacol and syringol, which are intermediate in the synthesis of pharmaceutical and polymeric compounds, or aimed at producing adhesives (Effendi et al. 2008).

In spite of the scarce scientific research aimed at determining the effects of its application (efficacy or toxicity), pyroligneous acid has been widely used in Asia as a home remedy for treating inflammatory diseases. According to Lee et al. (2011), pyroligneous acid obtained from oak tree charcoal production has been used to alleviate cellulitis, athlete's foot, and atopic dermatitis symptoms. In Japan, pyroligneous acid obtained from bamboo has long been used for the alternative treatment of eczema, scabies, atopic dermatitis, and other skin diseases. Studies by Kimura et al. (2002) about the carcinogenicity of commercialized pyroligneous acid from bamboo had suggested that it has no carcinogenic effect and do not promote tumors. Ho et al. (2013) have successfully demonstrated the anti-inflammatory activity of pyroligneous acid from bamboo both in vitro and in vivo by decreasing the expression of inflammatory mediators including nitric oxide (NO) generation, iNOS expression, and IL-6 expression. According to their study, creosol (2-methoxy-4-methylphenol) found in phenolic fraction of pyroligneous acid has been identified as the main compound with anti-inflammatory activity.

In a previous study, we explored the potential of pyroligneous acid from PKS as anti-inflammatory agent (Mahmud 2017). Results indicate that it shows to inhibit the generation of free radical nitric oxide (NO) in the LPS-activated RAW264.7 cell in a dose-dependent manner. In the presence of 25 μg/mL of pyroligneous acid, the inhibition of NO production was 93.45%, and the IC_{50} value was in the range of 3.125–6.25 μg/mL. These results show that pyroligneous acid from PKS has a good performance in inhibiting NO production. In addition, the cell viability assay did not reveal any significant cytotoxicity effect on cells treated with pyroligneous acid at doses up to 50 μg/mL. A high content of phenols (83%wt.) was identified in the compositional analysis of the pyroligneous acid from PKS. The major compounds were catechol, syringol, and 3-methoxycatechol. Other compounds identified were ketones and esters. It can be deduced that phenolic derivatives such as catechol and syringol play a major role in anti-inflammatory activity of this pyroligneous acid.

Pyroligneous acid had been successfully used as a food additive to impart flavor, color, and aroma to foods since 1880s. Smoke flavors are generally recognized as safe (GRAS), so they can be used as food preservative to prevent microbial growth provided their concentrations do not compromise the good manufacturing practice.

Pyroligneous acids contain bioactive compounds that have synergistic effects over the antimicrobial agents' action (Hwang et al. 2005; Darah et al. 2014). Bamboo pyroligneous acid has a significant effect against pathogenic bacteria such as *Escherichia coli*, *Staphylococcus pseudintermedius*, and *Pseudomonas aeruginosa* isolated from cats and dogs, with maximum inhibitory dilution (MID) of 109.4 ± 3.64 (100–120), 103.4 ± 7.41 (100–150), and 147.2 ± 8.87 (140–160), respectively. This pyroligneous acid has a greater efficacy against *P. aeruginosa* infections as compared to *E.coli and S. pseudintermedius*.

The so-called liquid smoke usually contains acetic acid which is responsible for its bacteriostatic properties (Holley and Patel 2005). It has been reported that the pyroligneous acid from *Rhizophora apiculata* contains 5.5%wt. acetic acid, 3.4%wt. methanol, and 6.5%wt. wood tar. The occurrence of high amounts of volatile compounds (8–10%wt.) contributes to the low pH of the pyroligneous acid (which is in the range of 2–3), as well as to its mild corrosive properties (Sipïla et al. 1998; Darah et al. 2014). Acetic acid can acidify the interior of bacterial cells resulting in degeneration and loss of bacterial components (Cherrington et al. 1991). It is thought that the high bactericidal activity of some liquid smokes is due to the high concentration of phenolic compounds in pyroligneous acid. This is because they have a high water solubility, and the polar phenolic compounds have greater opportunity to contact and interact with target organisms and exert higher levels of bacterial extermination. Cowan (1999) suggested that the group of phenolic derivatives is one of the responsible for antimicrobial activity of pyroligneous acid. The antibacterial activity of pyroligneous acid extracts may be due to the presence of lignans, flavonoids, astragalin, triterpenoids, glycosides, and tannins in the extracts (Darah et al. 2014). The high amounts of organic acids, phenolic compounds, and carbonyls have been reported to be responsible for the strong anti-biological activity (Vitt et al. 2001; Sameshima et al. 2002; Yatagai et al. 2002; Mun and Ku 2010; Wei et al. 2010; Zhai et al. 2015). Combinations of organic acid and phenolic have been reported by Darah et al. (2014) to result in a strong bactericidal inhibition. Phenolic derivatives are believed to be responsible for the antimicrobial activity, among which phenolic acids with low boiling points are good as bacteriostatic agents. Phenolic compounds such as 4-ethyl-2-methoxyphenol and 4-propyl-2-methoxyphenol have been identified as being primarily responsible for any antimicrobial activity (Lee et al. 2010). In addition, as temperature increases, the effectiveness of pyroligneous acid as anti-fungus and antibacterial agent increases (Wei et al. 2010; Lee et al. 2010). Then, phenolic fractions were used in food industries as antioxidant and antibacterial agents (Esekhiagbe et al. 2009; Achmadi et al. 2013).

Phenols inhibit the synthesis of nucleic acids of both Gram-negative and Gram-positive bacteria (Cushnie and Lamb 2005; Borges et al. 2012). A number of researchers have reported the relationship between the high concentration of phenols and the high inhibitory effects (Cowan 1999; Borges et al. 2012; Samy and Gopalakrishnakone 2010). Phenolic acids contain two distinctive carbon frameworks, the hydroxycinnamic and hydroxybenzoic structures. Although the basic skeleton remains the same, the numbers and positions of the hydroxyl groups on the aromatic ring causes significant changes in the properties of the phenolic products (Robbins 2003; Stalikas 2007). The number of hydroxyl groups and their position on the phenolic ring are thought to be related to their relative toxicity to microorganisms. Since, the higher the hydroxylation degree, the higher the toxicity (Cowan 1999; Samy and Gopalakrishnakone 2010), and evidence exists that increased hydroxylation results in increased toxicity (Cowan 1999; Samy and Gopalakrishnakone 2010) (Fig. 2.2).

Studies by Borges et al. (2012) revealed that gallic acid is capable to reduce 80% of biofilm form by *P. aeruginosa*, followed by *S. aureus* (70%) and *E. coli* (<70%). Those studies also reported about the effectiveness of phenolic compounds in comparison to ferulic acid as anti-biofilm agent. Jagani et al. (2009) studied the effect of phenolic compounds on biofilm formation by *P. aeruginosa*. Phenol was capable to reduce 80.9% of biofilm formation by *P. aeruginosa* followed by tannic acid (78.1% reduction). Polyphenol, polyanacardic acid, catechin, and epigallocatechin successfully reduced more than 60% of biofilm formation by *P. aeruginosa*. Ethyl linoleate and ascorbic acid are not effective in reducing *P. aeruginosa* biofilm. Extracellular polymeric substances (EPS) are resistant to antimicrobial agents as it forms a hydrated barrier between cells and their external environment (Mohamed et al. 2013). Individual bacterial strain produced higher EPS concentration as compared to bacteria grown in a noncompetitive environment (Subramaniam et al. 2010). Nikolic

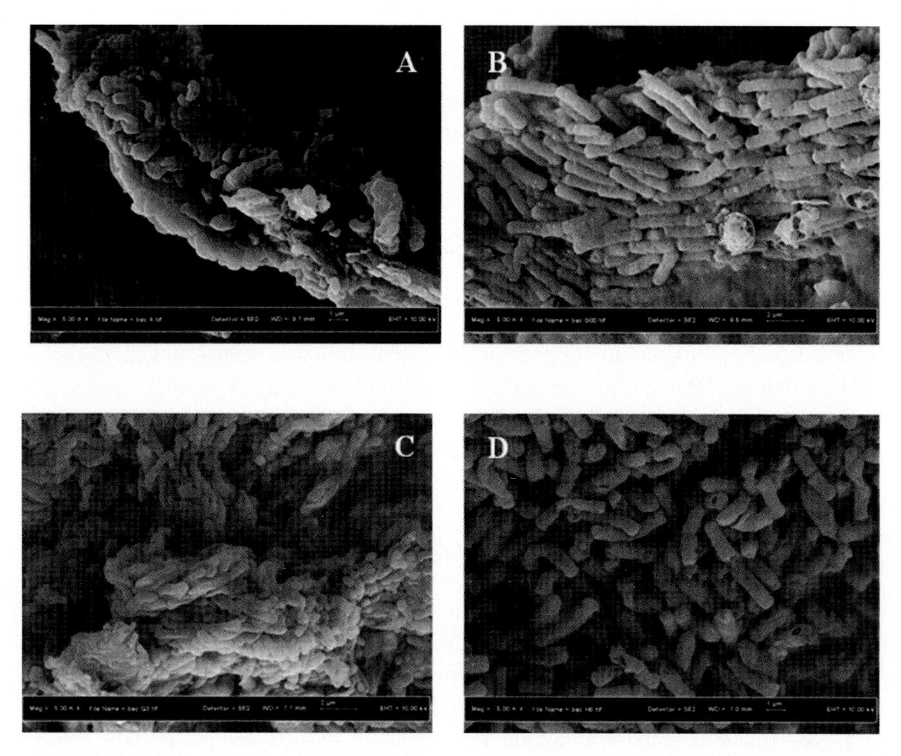

Fig. 2.2 Field emission scanning electron microscope micrographs for *B. subtilis* (**a**) without the addition of PA extract, (**b**) with the addition of PA extract and for *P. aeruginosa*, (**c**) without the addition of PA extract, and (**d**) with the addition of PA extract (magnification 5000×, bar – 2 μm)

et al. (2014) indicated that phenolic compounds from Z. officinale are able to perform as quorum sensing inhibitors, thus reducing the formation of biofilm. Another study by Siddique et al. (2012)) on the effect of *Piper betle* extract (PBE) over *P. aeruginosa* showed that EPS production by *P. aeruginosa* was reduced after treatment with PBE including phenolic compounds. Carvacrol extracted from plants also has an ability in reducing EPS production as described by Neyret et al. (2014), where the EPS production of the bacteria was reduced upon contact with carvacrol for 24 h, as it was evident from the clearer appearance of bacterial cells as a whole unit.

2.4 Conclusions

The pyrolysis of residues from palm oil industry yielded almost 35%wt. of liquid products in the case of the palm kernel shell and 39–46%wt. for the case of different empty fruit bunches. The by-products obtained, i.e., char (HHV 18.3 MJ kg-1, 52%wt. yield in the pyrolysis of PKS and 33–41%wt. yield in the case of different EFB) and gases (HHV 5.9 MJ kg-1, 13%wt. yield in the pyrolysis of PKS and

19–23%wt. in the case of EFB samples) could be used as energy sources in self-sustaining large-scale processes which can produce liquid products aimed at obtaining chemicals from the liquid fractions. In the pyrolysis of palm kernel shell, the liquid fractions (pyroligneous acid and bio-oil) had a high content of acetic acid and phenolic compounds (mainly phenol) and significant proportions of methanol and furfural. All these products are useful chemicals with numerous industrial applications and, most importantly, came from a renewable source.

Acknowledgment This work was performed with the financial assistance of Universidad Nacional del Litoral (Santa Fe, Argentina), Secretary of Science and Technology, Proj. CAID 2011 #501-201101-00546LI; Consejo Nacional de Investigaciones Científicas y Técnicas (CONICET, PIP 593/13), and the CONICET-UTM bilateral international cooperation project. The authors also acknowledge Universiti Teknologi Malaysia (UTM) for the Flagship Grant (03G44) and Postdoctoral Fellowship Scheme to Teo Chee Loong, Ministry of Education Malaysia for the MyPhD scholarship to Khoirun Nisa Mahmud, and Public Service Department for the King's scholarship to Suzami Junaidah Ariffin.

References

Achmadi S, Mubarik N, Nursyamsi R, Septiaji P (2013) Characterization of redistilled liquid smoke of oil-palm shells and its application as fish preservatives. J Appl Sci 13:401–408

Akhtar J, Idris A, Abd AR (2014a) Recent advances in production of succinic acid from lignocellulosic biomass. Appl Microb Biotechnol 98:987–1000

Akhtar J, Idris A, Teo C, Lai L, Hassan N, Infran kan M (2014b) Comparison of delignification of oil palm empty fruit bunch (EFB) by microwave assisted alkali/acid pretreatment and conventional pretreatment method. Int J Adv Chem Eng Biol Sc 1:155–157

Alén R, Kuoppala E, Oesch P (1996) Formation of the main degradation compound groups from wood and its components during pyrolysis. J Anal Appl Pyrolysis 36:137–148

Ani F (2015) Microwave processing of agricultural waste biomass to bio-oil. In: Idris A, Sedran U, Zakaria Z (eds) Biotechnology development in agriculture, industry and health. Advanced conversion technologies for lignocellulosic biomass, vol 3. Penerbit UTM Press, Malaysia, pp 67–93

Bayerbach R, Meier D (2009) Characterization of the water-insoluble fraction from fast pyrolysis liquids (pyrolytic lignin). Part IV: structure elucidation of oligomeric molecules. J Anal Appl Pyrolysis 85:98–107

Bertero M, Sedran U (2014) Chapter 14: co-processing of bio-oil in FCC. In: Pandey A (ed) Thermochemical conversion of biomass. Elsevier

Bertero M, de la Puente G, Sedran U (2011) Effect of pyrolysis temperature and thermal conditioning on the coke-forming potential of bio-oils. Energy Fuel 25:1267–1275

Bertero M, de la Puente G, Sedran U (2012) Fuels from bio-oils: bio-oil production from different residual sources, characterization and thermal conditioning. Fuel 95:263–271

Bertero M, Gorostegui H, Orrabalis C, Guzmán C, Calandri E, Sedran U (2014) Characterization of the liquid products in the pyrolysis of residual chañar and palm fruit biomasses. Fuel 116:409–414

Bhatia L, Johri S, Ahmad R (2012) An economic and ecological perspective of ethanol production from renewable agro waste: a review. AMB Express 2:1–19

Borges A, Maria J, Simoes M (2012) The activity of ferulic and gallic acids in biofilm prevention and control of pathogenic bacteria. Biofouling: J Bioadhesion Biofilm Res 28:755–767

Cherrington C, Hinton M, Mead G, Chopra I (1991) Organic acids: chemistry, antibacterial activity and practical applications. Adv Microb Physiol 32:87–108
Chin M, Poh P, Tey B, Chan E, Chin K (2013) Biogas from palm oil mill effluent (POME): opportunities and challenges from Malaysia's perspective. Renew Sust Energ Rev 26:717–726
Collard F, Blin (2014) J Renew. Sust Energy Rev 38:594–608
Cowan M (1999) Plant products as antimicrobial agents. Clin Microbiol Rev 12:564–582
Cushnie T, Lamb A (2005) Antimicrobial activity of flavonoids. Int J Antimicrob Agents 26:343–356
Darah I, Kassim J, Lim S, Rusli W (2014) Evaluation of antibacterial effects of Rhizophora apiculata pyroligneous acid on pathogenic bacteria. Malays J Microbiol 10:197–204
Demirbas A (2001) Carbonization ranking of selected biomass for charcoal, liquid and gaseous product. Energy Convers Manag 42:1229–1238
Dit M (2007) Palm Kernel Shell (PKS) is more than biomass for alternative fuel after 2005. Extract of PIPOC 2007: proceedings of chemistry and technology conference
Dutta S, De S, Alam I, Abu-Omar M, Saha B (2012) Direct conversion of cellulose and lignocellulosic biomass into chemicals and biofuel with metal chloride catalysts. J Catal 288:8–15
Effendi A, Gerhauser H, Bridgwater A (2008) Production of renewable phenolic resins by thermochemical conversion of biomass: a review. Renew Sust Energ Rev 12:2092–2116
Egsgaard H, Larsen E (2000) Thermal transformation of light tar. Specific routes to aromatic aldehydes and PAH. 1st world conference on biomass form energy and industry, Sevilla, Spain, pp 1468–1474
Esekhiagbe M, Agatemor M, Agatemor C (2009) Phenolic contents and antimicrobial potentials of Xylopia aethiopica and Myristica argentea. Maced J Chem Eng 28:159–162
Freel B, Graham R. (2002) U.S. Patent 6,485-841
Hamzah F, Idris A, Rashid R, Ming S (2009) Lactic acid production from microwave-alkali pretreated empty fruit bunches fibre using Rhizopus oryzae pellet. J Apll Science 9(17):3086–3091
Ho C, Lin C, Ka S, Chen A, Tasi Y, Liu M (2013) Bamboo vinegar decreases inflammatory mediator expression and NLRP3 inflammasome activation by inhibiting reactive oxygen species generation and protein kinase C-α/δ activation. PLoS One 8:1–11
Hoe T, Sarmidi M, Alwee S, Zakaria Z (2016) Recycling of oil palm empty fruit bunch as potential carrier for biofertilizer formulation. Journal Teknologi Sciences Engineering 78:165–170
Holley R, Patel D (2005) Review. Improvement in shelf-life and safety of perishable foods by plant essential oils and smoke antimicrobials. Food Microbiol 22:273–292
Hosoya T, Kawamoto H, Saka S (2007) Secondary reactions of lignin-derived primary tar components. Anal Appl Pyrolysis 80:118–125
Hu Z, Wang Y, Wen Z (2008) Alkali (NAOH) pre-treatment of switchgrass by radio frequency-based dielectric heating. Appl Biochem Biotechnol 148:71–81
Hwang Y, Matsushita Y, Sugamoto K, Matsui T (2005) Antimicrobial effect of the wood vinegar from Crytomeria japonica sapwood on plant pathogenic microorganisms. J Microbiol Biotechnol 15:1106–1109
Jagani S, Chelikani R, Kim D (2009) Effects of phenol and natural phenolic compounds on biofilm formation by Pseudomonas aeruginosa. Biofouling 25:321–324
Kawser M, Farid Nash A (2000) Oil palm shell as a source of phenol. J Oil Palm Res 12:86–94
Kim S, Jung S, Kim J (2010) Fast pyrolysis of palm kernel shells: influence of operation parameters on the bio-oil yield and the yield of phenol and phenolic compounds. Bioresour Technol 101(23):9294–9300
Kimura Y, Suto S, Tatsuka M (2002) Evaluation of carcinogenic/co-carcinogenic activity of chikusaku-eki, a bamboo charcoal by-product used as a folk remedy, in BALB/c3T3 cells. Biol Pharm Bull 25:1026–1029
Lee S, Lee A, Sajap N, Tey B (2010) Production of pyroligneous acid from lignocellulosic biomass and their effectiveness against biological attacks. J Appl Sci 10:2440–2446

Lee C, Yi E, Kim H, Huh S, Sung S, Chung M, Ye S (2011) Anti-dermatitis effects of oak wood vinegar on the DNCB-induced contact hypersensitivity via STAT3 suppression. J Ethnopharmacol 135:747–753

Lin Y, Cho J, Tompsett G, Westmoreland P, Huber G (2009) Kinetics and mechanism of cellulose pyrolysis. J Phys Chem C 113:20097–20107

Loo A, Jain K, Darah I (2007) Antioxidant activity of compounds isolated from the PA rhizophora apiculata. Food Chem 104:300–307

Mahmud, K (2017) Optimization of total phenolic contents in pyroligneous acid from oil palm kernel shell and its bioactivities, Ph.D thesis, Universiti Teknologi Malaysia

Malaysia Palm Oil Board (MPOB) (2015) Economics & Industry Development Division, MPOB Area 2015

Mathew S, Zakaria A, Musa N (2015) Antioxidant property and chemical profile of pyroligneous acid from pineapple plant waste biomass. Process Biochem 50:1985–1992

Mohamed A, Hamzah Z, Mohamed Daud M, Zakaria Z (2013) The effects of holding time and the sweeping nitrogen gas flowrates on the pyrolysis of EFB using a fixed bed reactor. Proc Eng 53:185–191

Mosier N, Wyman C, Dale B, Elander R, Lee Y, Holtzapple M, Ladisch M (2005) Features if promising technologies for pre-treatment of lignocellulosic biomass. Bioresour Technol 96:673–686

Mun S, Ku C (2010) Pyrolysis GC-MS analysis of tars formed during the aging of wood and bamboo crude vinegars. J Wood Sci 56:47–52

Neyret C, Herry H, Meylheuc T, Brissonnet F (2014) Plants derived compounds as natural antimicrobials paper mill biofilms. J Ind Microbiol Biotechnol 41:87–96

Nikolic M, Vasic S, Durdevic J, Stefanovic O, Comic L (2014) Antibacterial and anti biofilm activity of ginger ethanolic extract Kragujevac. J Forensic Dent Sci 36:129–136

Özbay M, Apaydin-Varol E, Uzun B, Putün A (2008) Characterization of bio-oil obtained from fruit pulp pyrolysis. Energy 33:1233–1240

Pütün A, Apaydin E, Pütün E (2002) Bio-oil production from pyrolysis and steam pyrolysis of soybean cake: products yields and composition. Energy 27:703–710

Pütün A, Apaydin-Varol E, Pütün E (2004) Rice straw as a bio-oil source via pyrolysis and steam pyrolysis. Energy 29:2171–2180

Pütün A, Uzun B, Apaydin-Varol E, Pütün E (2005) Bio-oil from olive oil industry wastes: pyrolysis of olive residue under different conditions. Fuel Process Technol 87:25–32

Robbins R (2003) Phenolic acids in foods: an overview of analytical methodology. J. Agric Food Chem 51:2866–2887

Sameshima K, Sasaki M, Sameshima I (2002) Fundamental evaluation on termiticidal activity of various liquid from charcoal making. Proceeding of the 4th international wood science symposium, Sept 2–5, Serpong, Indonesia, pp 134–138

Samy R, Gopalakrishnakone P (2010) Therapeutic potential of plants as anti-microbials for drug discovery. Evid Based Complement Alternat Med 7:283–294

Siddique M, Sakinah M, Ismail A, Matsuura T, Zularisam A (2012) The anti-biofouling effects of piper betle extracts against Pseudomonas aeruginosa and bacterial consortium. Desalination 288:24–30

Sipila K, Kuoppala E, Fagernäs L, Oasmaa A (1998) Characterization of biomass based flash pyrolysis oils. Biomass Bioenergy 14:103–113

Stalikas C (2007) Extraction, separation, and detection methods for phenolic acids and flavonoids. J Sep Sci 30:3268–3295

Stefanidis S, Kalogiannis K, Iliopoulou E, Michailof C, Pilavachi P, Lappas A (2014) A study of lignocellulosic biomass pyrolysis via the pyrolysis of cellulose, hemicellulose and lignin. J Anal Appl Pyrolysis 105:143–150

Subramaniam S, Yan S, Tyagi R, Surampalli R (2010) Extracellular polymeric substances (EPS) producing bacterial strains of municipal wastewater sludge: isolations, molecular identifications. Water Res 44:2253–2266

Sun Y, Cheng J (2002) Hydrolysis of lignocellulosic materials for ethanol production: a review. Bioresour Technol 83:1–11

Valle B, Castaño P, Olazar M, Bilbao J, Gayubo A (2012) Deactivating species in the transformation of crude bio-oil with methanol into hydrocarbons on a HZSM-5 catalyst. J Catal 285:304–314

Vitt S, Himelbloom B, Crapo C (2001) Pyroligneous acid. J Food Saf 2:111–125

Wei Q, Ma X, Dong J (2010) Preparation, chemical constituents and antimicrobial activity of pyroligneous acids from walnut tree branches. J Anal Appl Pyrolysis 87:24–28

Williams P, Besler S (1996) The influence of temperature and heating rate on the slow pyrolysis of biomass. Renew Energy 7:233–250

Wititsiri S (2011) Production of Pyroligneous acids from coconut shells and additional materials for control of termite workers, Odontotermes sp. and Striped Mealy Bugs, Ferrisia virgata. Songklanakarin. J Sci Technol 33:349–354

Yatagai M, Nishimoto M, Hori K, Ohira T, Shibata A (2002) Termiticital activity of wood vinegar, their components and their homologues. J Wood Sci 48:338–342

Zhai M, Shi G, Wang Y, Mao G, Wang D, Wang Z (2015) Chemical composition and biological activities of pyroligneous acid from walnut shell. BioResources 10:1715–1729

Zhu S, Wu Y, Yu Z, Wang C, Yu F, Jin S, Ding Y, Chi R, Liao J, Zhang Y (2006) Comparison of three microwave/chemical pre-treatment processes for enzymatic hydrolysis of rice straw. Biosyst Eng 34:279–283

Chapter 3
Application of Novel Biochars from Maize Straw Mixed with Fermentation Wastewater for Soil Health

Yuan Zhou, Yajun Tian, Liqiu Zhang, Yongze Liu, and Li Feng

Abstract Recently more and more researches have focused on the preparation of novel biochars for specific use in soil amendment. A series of novel biochars (MS) produced by maize straw mixed with different fermentation wastewater are introduced for their preparation and application for soil health. Preparation methods of novel biochars include physical activation, chemical activation, and blending modification. Physical activations are more efficient than chemical activations in enhancing pristine biochar's surface structure, while the chemical activations are more capable in creating special functional groups. Blending modification method, mixing different kinds of additives with waste biomass together before pyrolysis, is usually used to increase the nutrient contents. The modified novel biochars have excellent properties such as high surface area and pore volume, rich functional groups, and high nutrient contents. The application of novel biochars to soil can improve soil fertility, promote plant growth, and increase crop yield. After the application of the novel MS biochars in soil, the contents of soil organic carbon and nitrogen were significantly increased. The addition of 5% novel biochar to soil showed the best performance for ryegrass growth and H_2O_2 enzymatic activity enhancement.

3.1 Introduction

Biochar is a carbon-rich product generated from the pyrolysis of waste biomass under an oxygen-limited condition at relatively low temperature (<700 °C) (Lehmann et al. 2011; Yuan et al. 2011; Khan et al. 2013; Wang et al. 2015a). Examples of waste biomass for biochar preparation include hard woods, rice husk,

Y. Zhou • Y. Tian • L. Zhang • Y. Liu • L. Feng (✉)
Engineering Research Center for Water Pollution Source Control and Eco-remediation, College of Environmental Science & Engineering, Beijing Forestry University, Beijing, People's Republic of China
e-mail: zhangliqiu@163.com

Z.A. Zakaria (ed.), *Sustainable Technologies for the Management of Agricultural Wastes*, Applied Environmental Science and Engineering for a Sustainable Future, https://doi.org/10.1007/978-981-10-5062-6_3

tea waste, cotton stalks, wheat straw, and animal litters (Ro et al. 2010; Liu et al. 2012; Li et al. 2014; Rajapaksha et al. 2014; Song et al. 2014; Wang et al. 2015b). With the development of pyrolysis techniques, biochar has been increasingly considered to be an emerging alternative for potential application in the agricultural and environmental protection, e.g., soil improvement, crop yield enhancement, carbon sequestration, and pollutants removal and control in soil (Ahmad et al. 2014; Li et al. 2014; Cha et al. 2016; Ding et al. 2016; Novak et al. 2016a; Rajapaksha et al. 2016). Microwave-activated biochar has been reported to reduce the toxicity of soil contaminated by industrial activity (Kołtowski et al. 2017). Onetime addition of biochar to agricultural soil could increase crop yield compared to the control group, which indicated the potential of novel biochars in increasing crop yield (Griffin et al. 2017). In spite of the potential advantages in many aspects, biochar also has some problems such as low sorption ability for heavy metals and organic contaminants when it was used for soil remediation (Beesley et al. 2011; Ahmad et al. 2014; Almaroai et al. 2013).

In general, the preparation methods of novel biochars include physical activation, chemical activation, and blending modification, which can improve the properties of biochar such as surface area, pore structure, functional groups, and nutrient contents (Li et al. 2014; Jing et al. 2014; Shen et al. 2015; Van Vinh et al. 2015; Cha et al. 2016). Biochar application to soil can effectively improve soil in its physical and chemical properties such as organic carbon and nitrogen contents, cation exchange capacity (CEC), pH, water holding capacity, and increase the activities of H_2O_2 enzymatic and soil microorganisms as well (Steiner et al. 2007; Yuan et al. 2011; Khan et al. 2013; Xu et al. 2014; Tang et al. 2015; Wang et al. 2015b).

3.2 Preparation and Characterization of Novel Biochars

3.2.1 Preparation of Novel Biochars

In general, different kinds of waste biomass such as hard woods, rice husk, tea waste, fruit peel, poultry litter, animal manures, and sewage sludge can be used to prepare novel biochars (Chen et al. 2011; Uchimiya et al. 2011a; Liu et al. 2011; Mandal et al. 2017). These biochars are often modified to increase their surface area and pore volume, to form additional and abundant functional groups, or to enhance the nutrient contents in biochars. Modification methods can be divided into three main categories, i.e., physical activation, chemical activation, and blending modification. Physical activation, also called gas activation, uses various gases such as steam and carbon dioxide for biochar activation. Physical activation can improve the physicochemical properties of novel biochars by removing the by-products of incomplete combustion, oxidizing the carbon surface, and further increasing the surface area of biochars (Ahmed et al. 2016; Cha et al. 2016). Physical activation methods are usually simple and economically feasible and more effective in increasing surface structure than chemical modification method. Li et al. (2014) reported that the surface area of novel biochar activated by heating method reached 11.5 times more than the pristine biochar. By contrast, the surface area of biochar activated by chemical modifications was only 0.009–0.63 times as pristine biochar. The representative chemical activation agents include alkalines (KOH, NaOH), acids (HNO_3, H_2SO_4, H_3PO_4), organic solvents (methanol), metal salts ($MgCl_2$), and magnetic oxides (Fe_3O_4). Although chemical activation has several drawbacks, such as apparatus corrosion and high cost, its activation effect in creating additional and abundant function groups is better than that of physical activation. Chemical oxidation using HNO_3, $KMnO_4$, or HNO_3/H_2SO_4 mixture can introduce acidic functional groups such as carboxylic, phenolic, and hydroxyl on the surface of biochars (Liu et al. 2012; Li et al. 2014; Rajapaksha et al. 2016).

As a newly developed blending modification method, novel biochars can be prepared by mixing different kinds of additives (e.g., clay mixture, sewage sludge, nutrient-contained wastewater) with waste biomass together before pyrolysis in order to solve the problem of nutrition deficiency or nutrition imbalance of biochars. Biochar modified by montmorillonite and kaolinite significantly increased the biochars' trace element contents such as Fe and Al compared to that of unmodified biochar (Yao et al. 2014). Fermentation wastewater produced from food-making process is rich in many nutrients and minerals (e.g., N, P, K, Fe, Mg) and can be used as a nutritional supplement (Watson et al. 2015; Wu et al. 2016). The addition of wastewater containing high contents of nitrogen and phosphorous to biochar presents a potential fertility and resource recycling opportunity. Thus, we tried to use maize straw mixed with different amounts of fermentation wastewater to prepare novel biochars from maize straw (MS) with high nutrient contents, and the preparation process is shown in Fig. 3.1.

Firstly, the dried maize straw powder was saturated with fermentation wastewater for 24 h, then placed in porcelain crucibles, covered with a magnetic crucible lid, and, finally, pyrolyzed in an intelligent box high-temperature muffle furnace (DC-B08/10) under 400 °C at a rate of 10 °C min^{-1}. Finally, the obtained product was washed by deionized water and oven-dried at 60 °C for 24 h. Samples are marked as MS0, MS1, MS2, and MS3 to represent biochars prepared from 0 to 3 ml/g wastewater dosage, respectively, and their effects on soil properties and plant growth are investigated in detail below. Different modification methods for novel biochars are as summarized in Table 3.1.

Fig. 3.1 The preparation process of novel maize straw (MS) biochars

Table 3.1 Physicochemical properties of novel biochars prepared by different activation methods

Modified methods of biochar	Feedstock	Temp (°C)	pH	CEC (cmol/kg)	Yield (%)	Surface area (m^2/g)	Pore volume (mL/g)	Elemental component (wt.%)							References
								C	H	O	N	C/N	O/C	H/C	
Without activation															
Without activation	Wheat straw	350	9.9	530	–	26.3	0.026	53.9	1.76	–	0.91	59.20	–	0.03	Koltowski et al. (2017)
Without activation	Bamboo	550	–	–	–	42.8	0.022	63.5	2.90	33.00	0.55	115.5	0.39	0.55	Li et al. (2014)
Without activation	Bur cucumber plants	300	10.9	–	51.83	0.85	0.004	65.9	5.55	23.09	5.08	13.00	0.35	0.08	Rajapaksha et al. (2015)
Without activation	Bur cucumber plants	700	12.3	–	27.52	2.3	0.008	69.4	1.31	24.45	4.61	15.06	0.35	0.02	
Without activation	*Miscanthus*	500	–	–	–	181.0	–	80.9	2.80	12.20	0.29	278.9	0.15	0.03	Shim et al. (2015)
Without activation	Hickory wood	600	–	45.7	–	256.0	–	84.7	1.83	11.30	0.30	282.3	0.13	0.02	Ding et al. (2016)
Without activation	Rice straw	450	10.4	–	53	522.5	1.200	44.2	2.24	–	1.44	30.69	–	0.05	Sherif and Elsherifb (2015)
Without activation	Corn stover	450	10.2	–	34	551.7	2.667	63.6	2.75	–	1.35	47.10	–	0.04	
Without activation	Bamboo	600	–	–	–	375.5	–	80.9	2.43	14.86	0.15	539.3	0.18	0.03	Yao et al. (2014)
Without activation	Bagasse	600	–	–	–	388.3	–	76.5	2.93	18.32	0.79	96.70	0.24	0.04	
Without activation	Hickory	600	–	–	–	401.0	–	81.8	2.17	14.02	0.73	112.1	0.17	0.03	
MS0	Maize straw	400	7.6	10.41	46.05	12.2	0.010	59.02	4.01	–	2.40	24.59	–	0.068	
Physical activation															
Steam	Cottonseed hull	650	9.9	–	25.4	34.0	–	91.0	1.50	5.90	1.60	56.87	0.03	0.02	Uchimiya et al. (2011a)

Steam	*Miscanthus*	500				332.0		82.1	2.67	11.00	0.31	264.8	0.13	0.03	Shim et al. (2015)
Steam	Turkey litter	700	–	–	39.9	66.7	–	44.8	0.91	5.80	1.94	23.08	0.13	0.02	Cantrell et al. (2012)
Steam	Wheat straw	350	8.8	–	–	246.2	0.159	–	–	–	–	–	–	–	Koltowski et al. (2017)
Steam	Bur cucumber plants	300	11.1	–	50.21	1.22	0.003	68.1	5.11	21.43	5.10	13.35	0.31	0.08	Rajapaksha et al. (2015)
Steam	Bur cucumber plants	700	11.7	–	18.9	7.10	0.038	50.6	1.66	44.88	2.54	19.90	0.89	0.03	
Heat	Bamboo	550	–	–	–	494.2	0.240	70.7	1.10	27.60	0.58	121.9	0.29	0.18	Li et al. (2014)
NH_3/CO_2	Cotton stalks	600	–	–	–	251.9	0.080	–	–	–	3.50	–	–	–	Xiong et al. (2013)
Chemical activation															
NaOH	Hickory wood	600	–	124.5	–	873	–	82.1	2.25	13.20	0.25	328.4	0.16	0.03	Ding et al. (2016)
$KMnO_4$	Bamboo	550	–	–	–	27.2	–	60.7	3.20	35.60	0.50	121.4	0.44	0.64	Li et al. (2014)
HNO_3	Bamboo	550	–	–	–	0.5	–	56.7	3.10	37.70	2.60	21.80	0.50	0.65	
NaOH	Bamboo	550	–	–	–	0.4	–	59.5	3.00	37.20	0.24	247.9	0.47	0.60	
KOH	Rice husk	500	7.0	–	–	117.8	0.073	76.4	3.30	0.90	16.9	4.50	0.01	0.04	Liu et al. (2012)
H_3PO_4	Rice straw	450	2.4	–	–	517.7	0.939	40.3	2.20	–	0.97	41.52	–	0.05	Sherif and Elsherifb (2015)
	Corn stover	450	2.4	–	–	513.9	0.646	55.2	2.45	–	1.08	51.10	–	0.04	
10% H_2SO_4	Rice husk	450	–	–	–	46.8	0.033	43.6	2.20	12.20	0.50	87.20	0.31	0.05	Liu et al. (2012)

(continued)

Table 3.1 (continued)

Modified methods of biochar	Feedstock	Temp (°C)	pH	CEC (cmol/kg)	Yield (%)	Surface area (m^2/g)	Pore volume (mL/g)	Elemental component (wt.%)							References
								C	H	O	N	C/N	O/C	H/C	
HNO_3	Rice husk	450	–	–	–	139.7	–	54.9	–	34.60	4.10	13.39	0.63	–	Liu et al. (2011)
Demineralization with HCl	Orange peel	150	–	–	82.4	22.8	0.023	50.6	6.20	41.00	1.80	28.11	0.79	0.06	Chen et al. (2011)
	Orange peel	200	–	–	61.6	7.8	0.010	57.9	5.50	34.40	1.90	30.47	0.59	0.09	
	Orange peel	250	–	–	48.3	33.3	0.020	65.1	5.10	26.50	2.20	29.59	0.41	0.08	
	Orange peel	300	–	–	37.2	32.3	0.031	69.3	4.50	22.20	2.40	28.88	0.32	0.20	
	Orange peel	350	–	–	33.0	51.0	0.010	73.2	4.20	18.30	2.30	31.83	0.25	0.05	
	Orange peel	400	–	–	30.0	34.0	0.010	71.7	3.50	20.80	1.92	37.34	0.29	0.05	
	Orange peel	500	–	–	26.9	42.4	0.019	71.4	2.30	20.30	1.83	39.02	0.28	0.03	
	Orange peel	600	–	–	26.7	7.8	0.008	77.8	2.00	14.40	1.80	43.22	0.19	0.02	
	Orange peel	700	–	–	22.2	201.0	0.035	71.6	1.70	22.20	1.50	47.73	0.31	0.03	
Demineralization with HCl	Pine needs	100	–	–	91.2	0.7	–	50.9	6.15	42.27	0.71	71.63	0.83	0.12	Chen et al. (2008)
	Pine needs	200	–	–	75.3	6.2	–	57.1	5.71	36.31	0.88	64.89	0.63	0.10	
	Pine needs	250	–	–	56.1	9.5	–	61.2	5.54	32.36	0.86	71.21	0.53	0.09	
	Pine needs	300	–	–	48.6	19.9	–	68.9	4.31	25.74	1.08	63.77	0.37	0.06	
	Pine needs	400	–	–	30.0	112.4	0.044	77.9	2.95	18.04	1.16	67.11	0.23	0.04	
	Pine needs	500	–	–	26.1	236.4	0.095	81.8	2.26	14.96	1.11	73.66	0.18	0.03	
	Pine needs	600	–	–	20.4	206.7	0.076	85.4	1.85	11.81	0.98	87.10	0.14	0.02	
	Pine needs	700	–	–	14.0	490.8	0.186	86.5	1.28	11.08	1.13	76.56	0.13	0.01	
Acidic aqueous	Broiler litter	350	–	–	–	60.0	0.000	45.6	4.00	18.30	4.50	10.13	0.41	0.09	Uchimiya et al. (2011b)
Acidic aqueous	Broiler litter	700	–	–	–	94.0	0.018	46	1.40	7.40	2.80	16.43	0.16	0.03	
$FeCl_3$	Wheat straw	450	8.3	–	–	50.0	0.038	25.9	1.70	21.60	0.60	43.20	0.83	0.07	Li et al. (2014)

Magnetite-Fe^{3+}/ Fe^{2+}	Orange peel	250	–	–	–	41.2	0.052	35.1	3.60	–	1.10	31.9	–	0.10	Chen et al. (2011)
	Orange peel	400	–	–	–	23.4	0.042	29.4	2.20	–	0.50	58.8	–	0.07	
	Orange peel	700	–	–	–	19.4	0.033	0.4	0.20	–	0.20	2.0	–	0.50	
KBr	Bamboo chips	600	9.6	–	28.1	246.7	2.621	81.2	2.83	8.27	4.55	17.85	0.10	0.03	Mandal et al. (2017)
	Corncob	600	10.1	–	29.7	242.1	5.536	79.1	2.87	8.86	4.25	18.61	0.10	0.03	
	Eucalyptus bark	600	9.4	–	28.9	188.1	5.575	79.1	3.30	12.17	4.20	18.83	0.15	0.04	
	Rice husk	600	9.3	–	35.2	159.1	6.121	57.2	2.28	7.13	3.96	14.44	0.12	0.04	
	Rice straw	600	8.8	–	38.3	220.2	19.230	49.5	2.56	9.24	5.16	9.59	0.19	0.05	
Mg loaded	Corncob	600	–	–	–	–	–	43.3	5.00	–	0.60	72.20	–	0.16	Fang et al. (2015)
Zn loaded	Pinecone	500	–	–	–	11.5	0.028	71.2	3.00	20.40	0.50	142.4	0.29	0.04	Van Vinh et al. (2015)
Methanol	Rice husk	450–500	–	–	–	66.0	0.051	71.3	3.60	23.40	0.80	89.10	0.32	0.05	Jing et al. (2014)
Amino	Sawdust chat	500	6.0	–	–	2.5	0.005	62.1	4.20	–	4.60	13.50	–	0.07	Yang and Jiang (2014)
Manganese oxide	Pine wood	600	–	–	–	463.1	0.022	80.0	1.90	14.60	0.30	266.7	0.18	0.02	Wang et al. (2015a)
MnOx	Corn straw	600	–	–	–	61.0	0.036	85.3	1.70	5.20	0.80	106.6	0.06	0.02	Song et al. (2014)

(continued)

Table 3.1 (continued)

Modified methods of biochar	Feedstock	Temp (°C)	pH	CEC (cmol/kg)	Yield (%)	Surface area (m^2/g)	Pore volume (mL/g)	Elemental component (wt.%)							References
								C	H	O	N	C/N	O/C	H/C	
Blending modification															
Clay mixture, montmorillonite	Bamboo	600	–	–	–	408.1	–	83.3	2.26	12.41	0.25	333.1	0.15	0.03	Yao et al. (2014)
	Bagasse	600	–	–	–	407.0	–	75.3	2.25	18.87	0.75	100.4	0.25	0.03	
	Hickory chips	600	–	–	–	376.1	–	80.9	2.21	15.14	0.28	289.0	0.19	0.03	
Clay mixture, kaolinite	Bamboo	600	–	–	–	239.8		81.0	2.15	15.85	0.25	324.1	0.20	0.03	
	Bagasse	600	–	–	–	328.6		70.2	2.44	24.44	0.74	94.9	0.35	0.03	
	Hickory chips	600	–	–	–	224.5		78.1	2.11	18.12	0.33	236.6	0.23	0.03	
Mineral addition	Rice straw	400	8.4	–	–	151.0	0.100	49.0	2.60	–	1.19	41.20	–	0.62	Li et al. (2014)
Mixed with swine solids	Chicken litter	620	–	–	43	–	–	41.5	1.20	0.7	2.80	14.82	0.02	0.04	Ro et al. (2010)
Fermentation wastewater-MS1	Maize straw	400	7.8	13.73	42.46	340.0	0.030	60.5	5.12	–	2.57	23.53	–	0.085	
Fermentation wastewater-MS2	Maize straw	400	8.0	14.82	34.49	366.0	0.060	61.8	4.58	–	2.66	23.24	–	0.074	
Fermentation wastewater-MS3	Maize straw	400	8.0	19.82	32.85	371.0	0.060	62.7	4.53	–	2.80	22.41	–	0.072	

3.2.2 *Characteristics of Novel Biochars*

3.2.2.1 Surface Area, Porosity, pH, and Elemental Compositions of Novel Biochars

Each kind of novel biochar has different physicochemical properties in terms of the surface area, pore volume, pH, and elementary composition (Table 3.1). The properties of novel biochars vary depending on the feedstock and modification methods, as well as the pyrolysis temperature (Chen et al. 2008; Chen and Chen 2009; Spokas et al. 2014a; Uchimiya et al. 2011a; Xiong et al. 2013; Yao et al. 2014). Generally, biochars modified by physical activations have higher pH value. For example, the pH value of a novel biochar made from bur cucumber plants reached 11.7 after steam activation under 700 °C, which directly suggests its useful application for neutralization of the acidic soil (Rajapaksha et al. 2015). In addition, some studies demonstrated that novel biochars prepared by blending modification have much higher nutrient contents and may further enhance the plant growth and crop yield (Ro et al. 2010; Li et al. 2014; Yao et al. 2014). As for the novel MS biochars, the nutrient contents, pH value, surface area, and pore volume increased proportionally with the increase in fermentation wastewater added. This, however, markedly reduced the yield of the biochars. The contents of total nitrogen of MS1, MS2, and MS3 were significantly higher compared to MS0, indicating that the addition of fermentation wastewater is beneficial for the improvement of nitrogen contents. The ratio of H/C, representing the degree of maturation and long-term stability of biochar (Spokas et al. 2014b), decreased from 0.085 to 0.072 with the increase of fermentation wastewater dosage from 1 to 3 ml/g. H/C values of MS biochars were all under the limit of 0.6 indicating the stability of biochars in soil (Schimmelpfennig and Glaser 2012). In addition, the CEC values increased from 10.41 to 19.82 cmol/kg with the increase of fermentation wastewater dosage from 0 to 3 ml/g. This directly indicates the significant role of MS biochar in the maintenance of the fertilizer maintenance and its buffering capacity, of which both factors are crucial for soil amendment.

3.2.3 *Functional Groups on the Surface of Novel Biochars*

Functional groups on the surface of biochars are primarily responsible for the adsorption ability toward heavy metal and organic contaminants. The type and quantity of functional groups of biochars vary greatly depending on their preparation methods and additives such as salts or oxides used for modifying the biochars (Li et al. 2014; Zhang et al. 2013; Shen et al. 2015). For example, biochar activated by microwave and steam led to the decomposition of bound hydrogen groups on the surface of biochar, but not for the aldehydes groups, which improved the removal of elemental mercury from medical residues compared to biochar without activation (Shen et al. 2015). Figure 3.2 shows the ATR-FTIR spectra results for raw maize

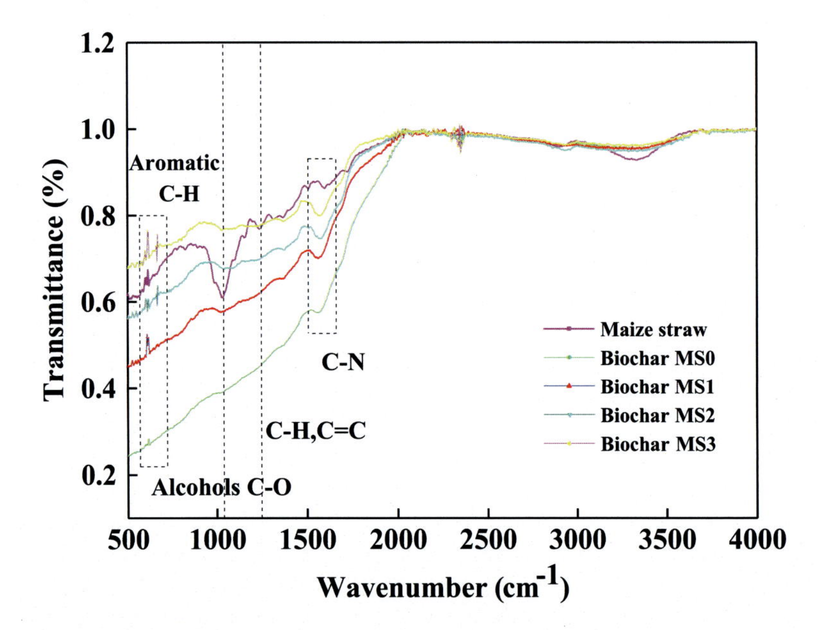

Fig. 3.2 ATR-FTIR spectra of MS biochars and maize straw

straw (MS) and after modification, i.e., MS biochars (MS0, MS1, MS2, MS3). Significant changes were observed between the intensities of bands of before and after modification steps. However, no noticeable difference was observed among the MS biochars produced. The C-O (1033 cm^{-1}) and C-H (1340–1465 cm^{-1}) groups in maize straw disappeared, while other functional groups such as aromatic C-H (600–1000 cm^{-1}) and C-N (1651 cm^{-1}) can be observed, and their peaks shifted to some extent. The disappearance of alcohol C-O bond indicated the decomposition of the initial constituents such as cellulose and lignin (Cantrell et al. 2012), while the shifts of C-N and aromatic C-H bonds indicate the structural modification during the preparation of biochars from maize straw.

3.3 Influences of Novel Biochars on Soil Health

In this study, the prepared MS2 biochar was added into soil with four mixing ratios (biochar to soil, 1%, 3%, 5%, and 10% w/w). The effect of MS2 biochar dosage on soil properties, plant growth (i.e., ryegrass growth), and H_2O_2 enzymatic activity by pot experiments was evaluated.

3.3.1 Influences of Novel Biochars on Soil Properties

The addition of biochar to soil could facilitate the stabilization of soil organic carbon and increase the holding capacity of soil for nitrogen, phosphorus, and potassium (Agegnehu et al. 2015; Kameyama et al. 2012). Biochars can provide soil with nitrogen, phosphorus, and potassium directly or can be used as nutrient supplement by microorganism activity indirectly (Steiner et al. 2008; Lehmann et al. 2011; Gul et al. 2015; Gul et al. 2016; Zhao et al. 2015; Ding et al. 2016). Moreover, the addition of biochar can also improve CEC of soil and retain non-acidic cations such as NH_4^+ and Mg^+ in soil, thus enhancing nutrient utilization efficiency (Agegnehu et al. 2015). In addition, biochar could also improve water retention capacity of soil, which is thought to be another important factor for plant growth (Shafie et al. 2012; Lim et al. 2016; Novak et al. 2016b). Shafie et al. found that water retention time of dry sandy soil is sharply increased from 5 to 15 days after the addition of biochars (Shafie et al. 2012).

In this study, MS2 biochar, maize straw biochar without modified plant ash, and organic fertilizer were applied in sandy soil to compare their influences on soil physicochemical properties. Table 3.2 shows physicochemical properties of soil after the addition of MS2 and other additives about 2 months. It can be observed that the pH, CEC, and nutrient contents in soil increased in different extent after the addition of MS2 and other additives, while the improvement of nutrient contents in the soil with MS2 biochar addition is more significant. For instance, the nitrogen content was increased by 109.8% after MS2 addition, which was much higher than that of other additives (34.1–78.0 %), the phosphorus content was increased by 23.1 % after MS2 addition and −34.6 to 19.2% for other additives, and the potassium content was increased by 131.8% and −10.0 to 127.3% for other additives.

3.3.2 Influences of Novel Biochars on Plant Growth and Crop Yield

Previous studies have demonstrated that biochars could promote plant growth and increase crop production (Asai et al. 2009; Gaskin et al. 2010; Agegnehu et al. 2015; Butnan et al. 2015). Agegnehu et al. reported that the addition of biochar including willow biochar and acacia biochar significantly increased maize height by 41.9–43.9% compared to the control group and supposed that the significant increase in maize growth may attribute to the raise of nutrient contents and water retention capacity (Agegnehu et al. 2015). Khan reported that the average shoot weight of rice was 11.1525 ± 0.8773 g in the control groups and increased to 19.10 ± 1.6658 g in sewage sludge biochar-amended soils (Khan et al. 2013). Other studies found that the response of crop yield to biochar application depends on several factors including the soil type, the biochar addition rate, and the kind of crops (Lentz and Ippolito 2012; Uzoma et al. 2011; Liu et al. 2013; Jones et al. 2012; Butnan et al. 2015; Ding et al.

Table 3.2 The influence of MS2 and other additives addition on soil physicochemical properties

Soil samples	pH	CEC (cmol/kg)	Organic carbon (g/kg)	Available N (g/kg)	Available P (g/kg)	Available K (g/kg)	Na (mg/kg)	Mg (mg/kg)	Fe (mg/kg)	Al (mg/kg)	Zn (mg/kg)
Sandy soil	6.42	14.82	17.01	0.41	0.26	0.22	0.36	1.78	1.46	10.79	0.04
5% maize straw biochar	6.93	32.81	25.25	0.58	0.31	0.24	1.28	2.09	0.73	34.93	3.53
5% novel biochar	7.19	39.85	39.28	0.86	0.32	0.51	1.75	4.75	1.99	37.04	4.06
5% plant ash	8.14	89.17	24.72	0.55	0.17	0.50	0.41	1.04	3.48	36.33	3.21
5% organic fertilizer	8.37	22.49	46.48	0.73	0.29	0.20	0.22	3.46	4.08	34.51	4.17

2016). For example, maize yield and water utilization efficiency increased by 98–150% and 91–139% as a result of biochar addition (Uzoma et al. 2011), and wheat grain yield increased by 18% due to the addition of oil mallee biochar (Solaiman et al. 2010).

Figure 3.3 shows the status of plant (ryegrass) growth after MS2 addition. It can be seen that when MS2 addition ratio was below 5%, MS2 could significantly stimulate plant growth compared to the control group and the leaf length and fresh weight of ryegrass reached the highest values of 12.31 cm and 5.75 g at 5% MS2 addition ratio. However, as MS2 addition ratio increased to 10%, the fresh weight of ryegrass was even lower than that in control group, which indicated that excessive MS2 addition might cause inhibition effects on ryegrass growth. This result may be attributed to the obstructed transportation of water, oxygen, and nutrients caused by harden soil as MS2 addition ratio was too high. Therefore, optimal MS2 addition proportion in this study was about 5%.

3.3.3 *Influences of Novel Biochars on Soil H_2O_2 Enzymatic Activity*

Biochar addition into soil has been found to increase the activities of many enzymes related to utilization of nutrients (Bailey et al. 2011; Liang et al. 2006). Many studies have mainly focused on extracellular enzymes activities after biochar applied to soil (Bailey et al. 2011; Wang et al. 2015b; Yang et al. 2016). However, very little is known about the variations of H_2O_2 enzyme activity, which can reflect the intensity of soil respiration and has often been used as an important indicator of soil fertilizer ability and situation of plant growth. In this study, the influence of MS2 addition on soil H_2O_2 enzymatic activity was investigated (Fig. 3.4).

It can be seen that H_2O_2 enzymatic activities in soils amended by four MS2 addition ratios all increased during the initial 45 days and then began to decrease. In the initial period after MS2 added to the soil, abundant nutrients promoted microbial activities and thus enhanced the H_2O_2 enzymatic activity. With the consumption of nutrients, microbial activities became to abate and caused the decreasing of H_2O_2 enzymatic activity. In this study, the H_2O_2 enzymatic activity in soil at 5% MS2 addition ratio reached the highest value of 2.93 mgH_2O_2/g soil after 45 days, which was in good agreement with the results of ryegrass growth. In other words, 5% MS2 addition ratio can not only enhance the ryegrass growth but also increase H_2O_2 enzymatic activity. It was also found that H_2O_2 enzymatic activities in MS2-amended soils except for 10% MS2 addition ratio were apparently higher than that in the maize straw biochar-amended soil.

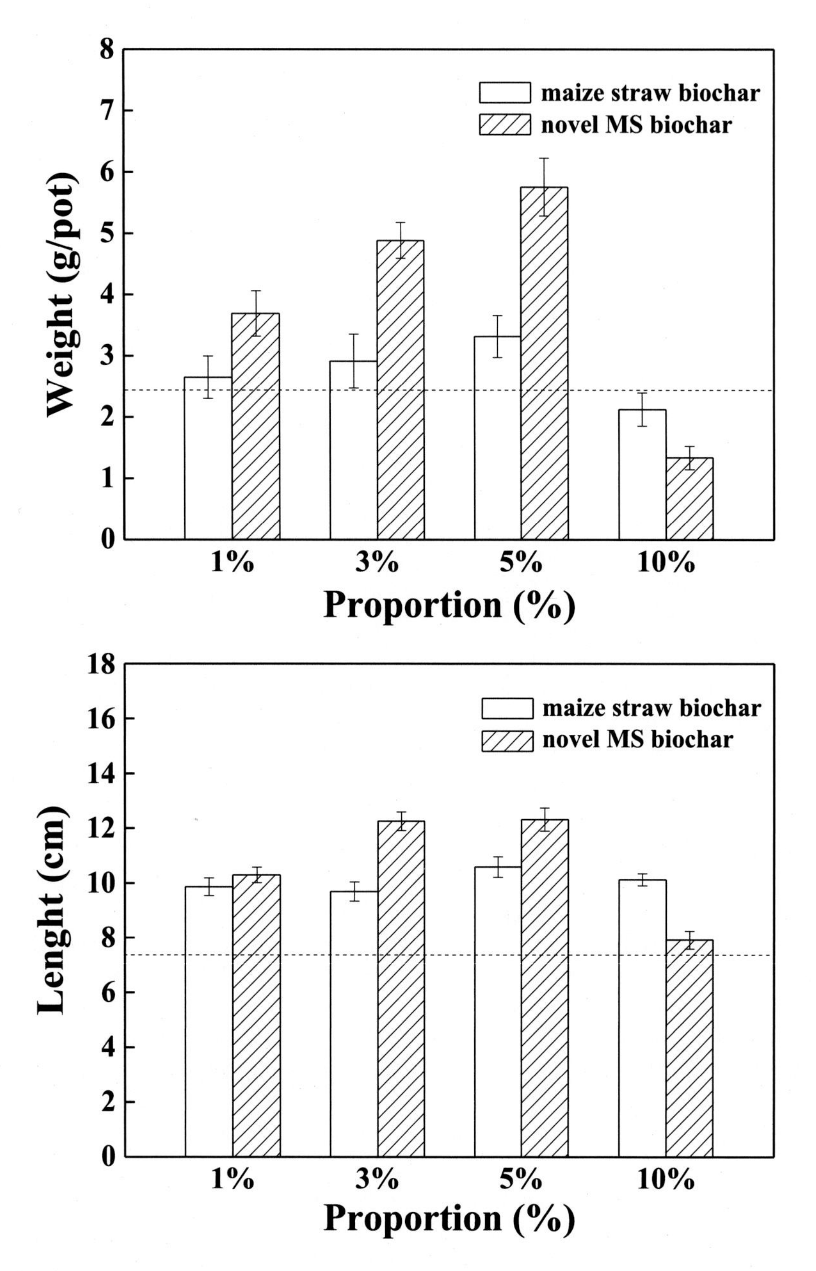

Fig. 3.3 The influences of MS2 addition ratio on ryegrass leaf length and fresh weight

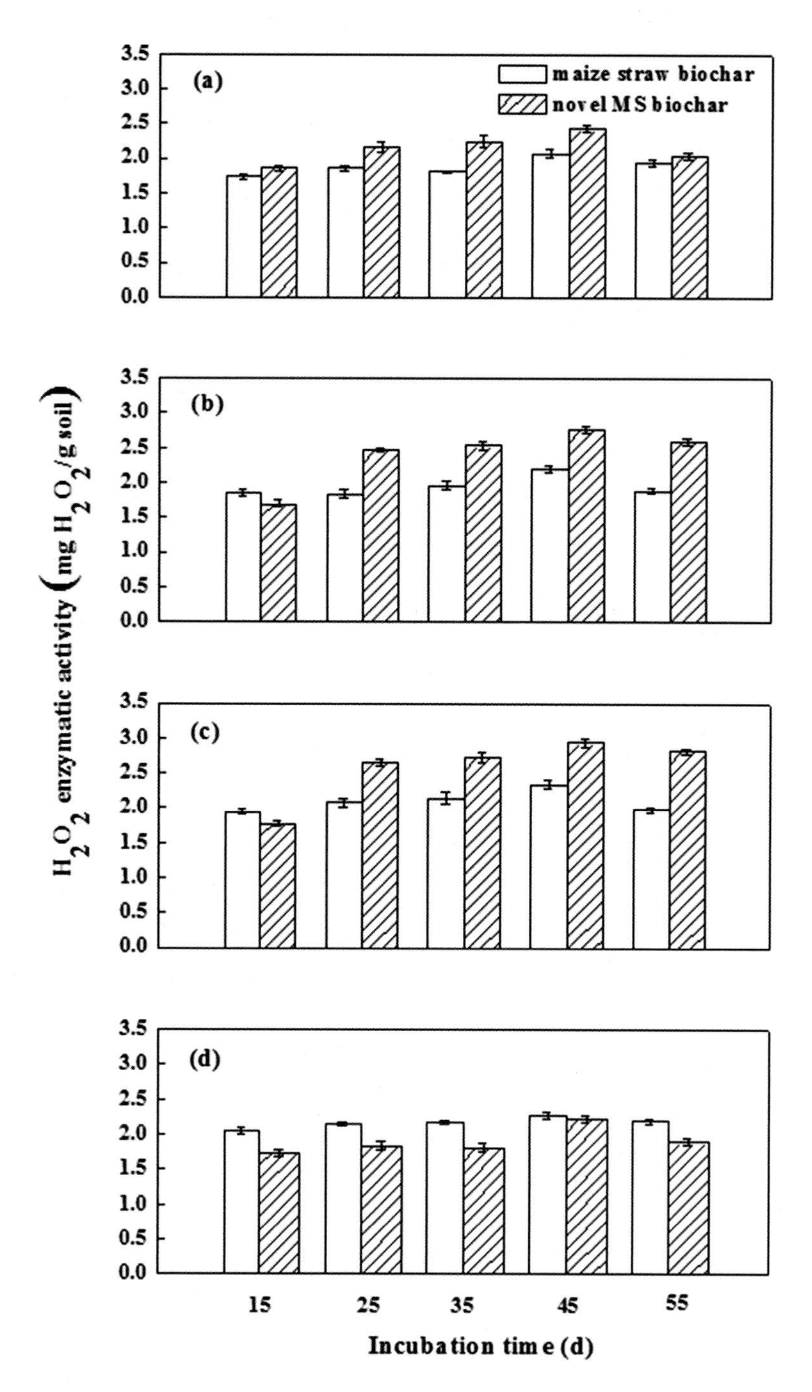

Fig. 3.4 The influence of different MS2 addition ratios on soil H_2O_2 enzymatic activity ((**a**) 1%, (**b**) 3%, (**c**) 5%, (**d**) 10%)

3.4 Future Prospect

Most of current studies are pot experiments and the experiment span is limited under 1 year. The long-term behavior of biochars added to soil and its potential negative impacts on soil quality are still uncertain. Therefore, it is urgent to investigate the long-term effects of biochars addition on soil quality and plant growth. Moreover, extensive research suggest that the applications of novel biochars to soil can improve soil fertility, promote plant growth, and increase crop yield, but the precise mechanism is still not clear.

3.5 Conclusion

This study summarizes different modification methods of novel biochars and highlights the preparation method of novel biochars by mixing maize straw with different amounts of fermentation wastewater and then investigates their influences on soil properties, plant growth, and H_2O_2 enzymatic activity. Fermentation wastewater addition is an effective means to increase organic carbon and nitrogen contents in novel biochar, thus enhancing corresponding nutrients in soil. After MS2 biochar applied to soil, it can not only increase the soil nutrient contents but also promote the plant growth and H_2O_2 enzymatic activity. The 5% MS2 addition ratio showed the optimal performance for ryegrass growth and H_2O_2 enzymatic activity enhancement.

Acknowledgment This work was supported by Education Committee of Beijing, China (2015GJ-02), and the Special S&T Project on Treatment and Control of Water Pollution (2013ZX07201007-003) for financial support.

References

Agegnehu G, Bass AM, Nelson PN, Muirhead B, Wright G, Bird MI (2015) Biochar and biochar-compost as soil amendments: effects on peanut yield, soil properties and greenhouse gas emissions in tropical North Queensland, Australia. Agric Ecosyst Environ 213:72–85

Ahmad M, Rajapaksha AU, Lim JU, Zhang M, Bolan N, Mohan D, Vithanage M, Lee SS, Ok YS, (2014) Biochar as a sorbent for contaminant management in soil and water: a review. Chemosphere 99:19–33

Ahmed MB, Zhou JL, Ngo HH, Guo W, Chen M (2016) Progress in the preparation and application of modified biochar for improved contaminant removal from water and wastewater. Bioresour Technol 214:836–851

Almaroai YA, Usman ARA, Ahmad M, Moon DH, Cho J-S, Joo YK, Jeon C, Lee SS, Ok YS (2013) Effects of biochar, cow bone, and eggshell on Pb availability to maize in contaminated soil irrigated with saline water. Environ Earth Sci 71(3):1289–1296

Asai H, Samson BK, Stephan HM, Songyikhangsuthor K, Homma K, Kiyono Y, Inoue Y, Shiraiwa T, Horie T (2009) Biochar amendment techniques for upland rice production in Northern Laos. Field Crop Res 111(1–2):81–84

Bailey VL, Fansler SJ, Smith JL, Bolton H (2011) Reconciling apparent variability in effects of biochar amendment on soil enzyme activities by assay optimization. Soil Biol Biochem 43(2):296–301

Beesley L, Moreno-Jimenez E, Gomez-Eyles JL, Harris E, Robinson B, Sizmur T (2011) A review of biochars' potential role in the remediation, revegetation and restoration of contaminated soils. Environ Pollut 159(12):3269–3282

Butnan S, Deenik JL, Toomsan B, Antal MJ, Vityakon P (2015) Biochar characteristics and application rates affecting corn growth and properties of soils contrasting in texture and mineralogy. Geoderma 237–238:105–116

Cantrell KB, Hunt PG, Uchimiya M, Novak JM, Ro KS (2012) Impact of pyrolysis temperature and manure source on physicochemical characteristics of biochar. Bioresour Technol 107:419–428

Cha JS, Park SH, Jung S-C, Ryu C, Jeon J-K, Shin M-C, Park Y-K (2016) Production and utilization of biochar: a review. J Ind Eng Chem 40:1–15

Chen B, Chen Z (2009) Sorption of naphthalene and 1-naphthol by biochars of orange peels with different pyrolytic temperatures. Chemosphere 76(1):127–133

Chen B, Zhou D, Zhu L (2008) Transitional adsorption and partition of nonpolar and polar aromatic contaminants by biochars of pine needles with different pyrolytic temperatures. Environ Sci Technol 42(14):5137–5143

Chen B, Chen Z, Lv S (2011) A novel magnetic biochar efficiently sorbs organic pollutants and phosphate. Bioresour Technol 102(2):716–723

Ding Y, Liu Y, Liu S, Li Z, Tan X, Huang X, Zeng G, Zhou L, Zheng B (2016) Biochar to improve soil fertility. a review. Agron Sustain Dev 36(2):36–36

Fang C, Zhang T, Li P, Jiang R, Wu S, Nie H, Wang Y (2015) Phosphorus recovery from biogas fermentation liquid by Ca–Mg loaded biochar. J Environ Sci 29:106–114

Gaskin JW, Speir RA, Harris K, Das KC, Lee RD, Morris LA, Fisher DS (2010) Effect of peanut hull and pine chip biochar on soil nutrients, corn nutrient status, and yield. Agron J 102(2):623–633

Griffin DE, Wang D, Parikh SJ, Scow KM (2017) Short-lived effects of walnut shell biochar on soils and crop yields in a long-term field experiment. Agric Ecosyst Environ 236:21–29

Gul S, Whalen JK (2016) Biochemical cycling of nitrogen and phosphorus in biochar-amended soils. Soil Biol Biochem 103:1–15

Gul S, Whalen JK, Thomas BW, Sachdeva V, Deng H (2015) Physico-chemical properties and microbial responses in biochar-amended soils: mechanisms and future directions. Agric Ecosyst Environ 206:46–59

Jing XR, Wang Y-Y, Liu WJ, Wang YK, Jiang H (2014) Enhanced adsorption performance of tetracycline in aqueous solutions by methanol-modified biochar. Chem Eng J 248:168–174

Jones DL, Rousk J, Edwards-Jones G, DeLuca TH, Murphy DV (2012) Biochar-mediated changes in soil quality and plant growth in a three year field trial. Soil Biol Biochem 45:113–124

Kameyama K, Miyamoto T, Shiono T (2012) Influence of sugarcane bagasse-derived biochar application on nitrate leaching in calcaric dark red soil. J Environ Qual 41(4):1131–1137

Khan S, Chao C, Waqas M, Arp HP, Zhu YG (2013) Sewage sludge biochar influence upon rice Oryza sativa L yield, metal bioaccumulation and greenhouse gas emissions from acidic paddy soil. Environ Sci Technol 47(15):8624–8632

Koltowski M, Charmas B, Skubiszewska-Zieba J, Oleszczuk P (2017) Effect of biochar activation by different methods on toxicity of soil contaminated by industrial activity. Ecotoxicol Environ Saf 136:119–125

Lehmann J, Rillig MC, Thies J, Masiello CA, Hockaday WC, Crowley D (2011) Biochar effects on soil biota – a review. Soil Biol Biochem 43(9):1812–1836

Lentz RD, Ippolito JA (2012) Biochar and manure affect calcareous soil and corn silage nutrient concentrations and uptake. J Environ Qual 41(4):1033–1043

Li Y, Shao J, Wang X, Deng Y, Yang H, Chen H (2014) Characterization of modified biochars derived from bamboo pyrolysis and their utilization for target component (furfural) adsorption. Energy Fuel 28(8):5119–5127
Liang B, Lehmann J, Solomon D, Kinyangi J, Grossman J, O'Neill B, Skjemstad JO, Thies J, Luizão FJ, Petersen J, Neves EG (2006) Black carbon increases cation exchange capacity in soils. Soil Sci Soc Am J 70(5):1719–1730
Lim TJ, Spokas KA, Feyereisen G, Novak JM (2016) Predicting the impact of biochar additions on soil hydraulic properties. Chemosphere 142:136–144
Liu WJ, Zeng FX, Jiang H, Zhang XS (2011) Preparation of high adsorption capacity bio-chars from waste biomass. Bioresour Technol 102(17):8247–8252
Liu P, Liu WJ, Jiang H, Chen JJ, Li WW, Yu HQ (2012) Modification of bio-char derived from fast pyrolysis of biomass and its application in removal of tetracycline from aqueous solution. Bioresour Technol 121:235–240
Liu X, Zhang A, Ji C, Joseph S, Bian R, Li L, Pan G, Paz-Ferreiro J (2013) Biochar's effect on crop productivity and the dependence on experimental conditions—a meta-analysis of literature data. Plant Soil 373(1–2):583–594
Mandal A, Singh N, Purakayastha TJ (2017) Characterization of pesticide sorption behaviour of slow pyrolysis biochars as low cost adsorbent for atrazine and imidacloprid removal. Sci Total Environ 577:376–385
Novak JM, Ippolito JA, Lentz RD, Spokas KA, Bolster CH, Sistani K, Trippe KM, Phillips CL, Johnson MG (2016a) Soil health, crop productivity, microbial transport, and mine spoil response to biochars. Bioenergy Res 9(2):454–464
Novak J, Sigua G, Watts D, Cantrell K, Shumaker P, Szogi A, Johnson MG, Spokas K (2016b) Biochars impact on water infiltration and water quality through a compacted subsoil layer. Chemosphere 142:160–167
Rajapaksha AU, Vithanage M, Zhang M, Ahmad M, Mohan D, Chang SX, Ok YS (2014) Pyrolysis condition affected sulfamethazine sorption by tea waste biochars. Bioresour Technol 166:303–308
Rajapaksha AU, Vithanage M, Ahmad M, Seo DC, Cho JS, Lee SE, Lee SS, Ok YS (2015) Enhanced sulfamethazine removal by steam-activated invasive plant-derived biochar. J Hazard Mater 290:43–50
Rajapaksha AU, Chen SS, Tsang DC, Zhang M, Vithanage M, Mandal S, Gao B, Bolan NS, Ok YS (2016) Engineered/designer biochar for contaminant removal/immobilization from soil and water: potential and implication of biochar modification. Chemosphere 148:276–291
Ro KS, Cantrell KB, Hunt PG (2010) High-temperature pyrolysis of blended animal manures for producing renewable energy and value-added biochar. Indust Eng Chem Res 49(20):10125–10131
Schimmelpfennig S, Glaser B (2012) One step forward toward characterization: some important material properties to distinguish biochars. J Environ Qual 41(4):1001
Shafie ST, Salleh MAM, Hang LL, Rahman MM, Ghani WAWAK (2012) Effect of pyrolysis temperature on the biochar nutrient and water retention capacity. J Purity Util React Environ 1(6):293–307
Shen B, Li G, Wang F, Wang Y, He C, Zhang M, Singh S (2015) Elemental mercury removal by the modified bio-char from medicinal residues. Chem Eng J 272:28–37
Sherif M, Elsherifb E (2015) Investigation of strontium (II) sorption kinetic and thermodynamic onto straw-derived biochar. Particulate Sci Technol 0:1–8
Shim T, Yoo J, Ryu C, Park YK, Jung J (2015) Effect of steam activation of biochar produced from a giant Miscanthus on copper sorption and toxicity. Bioresour Technol 197:85–90
Solaiman ZM, Blackwell P, Abbott LK (2010) Direct and residual effect of biochar application on mycorrhizal root colonisation, growth and nutrition of wheat. Soil Res 48(7):546–554
Song XD, Xue XY, Chen DZ, He PJ, Dai XH (2014) Application of biochar from sewage sludge to plant cultivation: influence of pyrolysis temperature and biochar-to-soil ratio on yield and heavy metal accumulation. Chemosphere 109:213–220

Spokas KA, Novak JM, Masiello A, Johnson G, Colosky EC, Ippolito JA, Trigo C (2014a) Physical disintegration of biochar: an overlooked process. Environ Sci Technol 1:326–332
Spokas KA (2014b) Review of the stability of biochar in soils: predictability of O:C molar ratios. Carbon Manag 1(2):289–303
Steiner C, Teixeira WG, Lehmann J, Nehls T, de Macêdo JLV, Blum WEH, Zech W (2007) Long term effects of manure, charcoal and mineral fertilization on crop production and fertility on a highly weathered Central Amazonian upland soil. Plant Soil 291(1–2):275–290
Steiner C, Glaser B, Geraldes Teixeira W, Lehmann J, Blum WEH, Zech W (2008) Nitrogen retention and plant uptake on a highly weathered central Amazonian Ferralsol amended with compost and charcoal. J Plant Nutr Soil Sci 171(6):893–899
Tang J, Lv H, Gong Y, Huang Y (2015) Preparation and characterization of a novel graphene/biochar composite for aqueous phenanthrene and mercury removal. Bioresour Technol 19:355–363
Uchimiya M, Chang S, Klasson KT (2011a) Screening biochars for heavy metal retention in soil: role of oxygen functional groups. J Hazard Mater 190(1–3):432–441
Uchimiya M, Klasson KT, Wartelle LH, Lima IM (2011b) Influence of soil properties on heavy metal sequestration by biochar amendment: 1. Copper sorption isotherms and the release of cations. Chemosphere 82(10):1431–1437
Uzoma KC, Inoue M, Andry H, Fujimaki H, Zahoor A, Nishihara E (2011) Effect of cow manure biochar on maize productivity under sandy soil condition. Soil Use Manag 27(2):205–212
Van Vinh N, Zafar M, Behera SK, Park HS (2015) Arsenic (III) removal from aqueous solution by raw and zinc-loaded pine cone biochar: equilibrium, kinetics, and thermodynamics studies. Int J Environ Sci Technol 12(4):1283–1294
Wang S, Gao B, Zimmerman AR, Li Y, Ma L, Harris WG, Migliaccio KW (2015a) Removal of arsenic by magnetic biochar prepared from pinewood and natural hematite. Bioresour Technol 175:391–395
Wang X, Zhou W, Liang G, Song D, Zhang X (2015b) Characteristics of maize biochar with different pyrolysis temperatures and its effects on organic carbon, nitrogen and enzymatic activities after addition to fluvo-aquic soil. Sci Total Environ 538:137–144
Watson VJ, Hatzell M, Logan BE (2015) Hydrogen production from continuous flow, microbial reverse-electrodialysis electrolysis cells treating fermentation wastewater. Bioresour Technol 195:51–66
Wu H, Lai C, Zeng G, Liang J, Chen J, Xu J, Dai J, Li X, Liu J, Chen M, Lu L, Hu L, Wan J (2016) The interactions of composting and biochar and their implications for soil amendment and pollution remediation: a review. Crit Rev Biotechnol:1–11
Xiong Z, Shihong Z, Haiping Y (2013) Influence of NH3/CO2 modification on the characteristic of biochar and the CO2 capture. Bioenergy Res 6(4):1147–1153
Xu HJ, Wang XH, Li H, Yao HY, Su JQ, Zhu YG (2014) Biochar impacts soil microbial community composition and nitrogen cycling in an acidic soil planted with rape. Environ Sci Technol 48(16):9391–9399
Yang GX, Jiang H (2014) Amino modification of biochar for enhanced adsorption of copper ions from synthetic wastewater. Water Res 48:396–405
Yang D, Yun GL, Sha BL (2016) Biochar to improve soil fertility. A review. Agron Sustain 36
Yao Y, Gao B, Fang J, Zhang M, Chen H, Zhoub Y, Creamer AE, Sun Y, Yang L (2014) Characterization and environmental applications of clay-biochar composites. Chem Eng J 242:136–143
Yuan JH, Xu RK, Zhang H (2011) The forms of alkalis in the biochar produced from crop residues at different temperatures. Bioresour Technol 102(3):3488–3497
Zhang M, Gao B, Varnoosfaderani S, Hebard A, Yao Y, Inyang M (2013) Preparation and characterization of a novel magnetic biochar for arsenic removal. Bioresour Technol 130:457–462
Zhao R, Coles N, Wu J (2015) Carbon mineralization following additions of fresh and aged biochar to an infertile soil. Catena 125:183–189

Chapter 4
Utilization of Oil Palm Fiber and Palm Kernel Shell in Various Applications

Maizatulakmal Yahayu, Fatimatul Zaharah Abas, Seri Elyanie Zulkifli, and Farid Nasir Ani

Abstract The agro-industrial sector in Malaysia generates a significant amount of biomass solid wastes, particularly from palm oil mills which produce enormous amount of biomass that includes empty fruit bunches (EFB), palm kernel shells (PKS), oil palm fibers (OPF), and palm oil mill effluents (POME). Oil palm fronds (leaves stem) are available from the plantation during trimming process, and oil palm trunks are available when the plantation are removing old palm trees for new oil palm cultivation. At present, the fibers and shells are utilized (in weight ratio) in boilers for steam generation in most palm oil mills, whereas empty fruit bunches are being utilized as organic fertilizer and mulching purposes. Thus, there are huge challenges as well as opportunities on the development to utilize these abundant biomass resources. The oil palm biomass appears to be the most promising and potential renewable feedstock for the recovery of various fuels and chemicals.

4.1 Introduction

Malaysia as a tropical country experiences hot and wet weather throughout the year. This climate encourages the growth of the oil palm and consequently the development of oil palm plantation in Malaysia. As the second largest producer of crude palm oil in the world, oil palm has become the fourth largest contributor to the country's gross national income (GNI) and thus generates around 80 million dry tons of biomass annually (Agensi Inovasi Malaysia 2012).

M. Yahayu (✉) • F.Z. Abas • S.E. Zulkifli
Institute of Bioproduct Development, Universiti Teknologi Malaysia,
Johor Bahru, Johor, Malaysia
e-mail: maizatul@ibd.utm.my

F.N. Ani
Department of Thermo-Fluids, Faculty of Mechanical Engineering, Universiti Teknologi Malaysia, Johor Bahru, Johor, Malaysia

Z.A. Zakaria (ed.), *Sustainable Technologies for the Management of Agricultural Wastes*, Applied Environmental Science and Engineering for a Sustainable Future, https://doi.org/10.1007/978-981-10-5062-6_4

Table 4.1 The oil palm biomass residue from palm oil industry (GGS 2013)

Oil palm waste biomass	Quantity (million tonnes/year)
Empty fruit bunches	21.90
Oil palm fibers	12.38
Oil palm shells	5.71
Palm oil mill effluents	55.22
Total	95.21

4.2 Biomass from Oil Palm Industry

The agro-industrial sector in Malaysia generates a significant amount of biomass solid wastes, particularly from palm oil mills which produces the enormous amount of biomass that includes empty fruit bunches (EFB), palm kernel shells (PKS), oil palm fibers (OPF), and palm oil mill effluents (POME). Oil palm fronds (leaves stem) are available from the plantation during trimming process, and oil palm trunks are available when the plantation are removing old palm trees for new oil palm cultivation. Malaysia has more than 400 palm oil mills and generated approximately 95.21 million tonnes of oil palm waste biomass in year-end 2012 (Table 4.1). At present, the fibers and shells are utilized (in weight ratio) in boilers for steam generation in most palm oil mills, whereas empty fruit bunches are being utilized as organic fertilizer and mulching purposes. Thus, there are huge challenges as well as opportunities on the development to utilize these abundant biomass resources. The oil palm biomass appears to be the most promising and potential renewable feedstock for the recovery of various fuels and chemicals.

4.3 Characteristics and Compositions of Oil Palm Fiber (OPF)

Fresh fruit bunches (FFB) are taken from the plantation and sent to the palm oil mill. The FFB are sterilized using steam to ease removal of the fruitlets from the FFB by stripping process. The fruitlets contain palm oil, fiber, and palm nuts. Oil palm fiber (OPF), which is also known as mesocarp fiber or pressed palm fiber, is one of the lignocellulosic biomass that consists of lignin, cellulose, and hemicelluloses. Palm oil is extracted from the pulp of the fruit (mesocarp). It is the residue that is obtained after pressing the mesocarp of palm fruitlets to obtain the palm oil. The cross section of the fruitlet is shown in Fig. 4.1a, and the waste OPF is shown in Fig. 4.1b.

The lignocellulosic content of OPF that has been reported by various researchers is given in Table 4.2. The OPF has higher amount of cellulose and hemicellulose content compared to lignin. During thermogravimetric analysis (TGA), the OPF will start to decompose at temperature of 150 °C, while at temperature of 300 °C, severe decomposition occurs on the internal structure of OPF (Sabil et al. 2013).

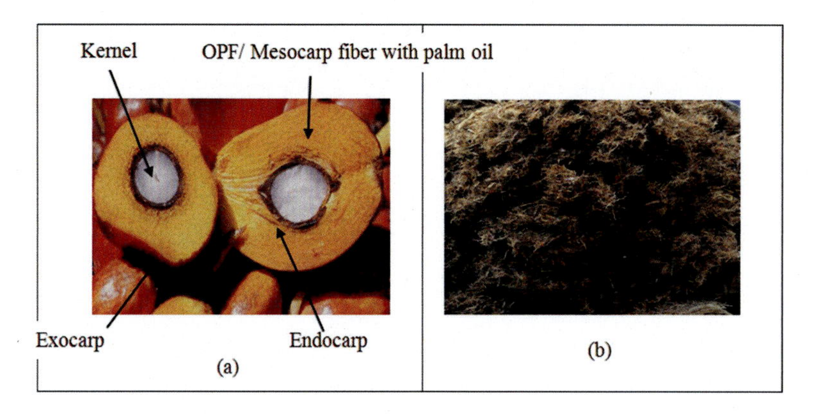

Fig. 4.1 (**a**) Cross section of oil palm fruitlet (Cookwithkathy 2013), (**b**) waste oil palm fiber

Table 4.2 Lignocellulosic content of oil palm fiber (OPF)

References	Lignin (%)	Cellulose (%)	Hemicellulose (%)
Chen and Lin (2016)	10.92	46.63	19.98
Abnisa et al. (2013)	27.3	23.7	30.5
Shinoj et al. (2011)	13–25	43–65	36–42

Table 4.3 Proximate and ultimate analysis of OPF

References	Chen and Lin (2016)	Aziz et al. (2011)	Abnisa et al. (2013)
Proximate analysis (%)			
Moisture content (%)	3.27	5.11	8.6
Volatile matter (%)	71.49	75.28	78
Fixed carbon (%)	19.18	14.63	7.6
Ash (%)	6.06	4.98	5.8
Ultimate analysis (wt %)			
Carbon	42.65	47.88	45.03
Oxygen	50.78	45.34	47.89
Nitrogen	1.09	0.63	0.94
Sulfur	–	0.41	–
Hydrogen	5.48	5.74	6.15

According to Shinoj et al. (2011), the diameter of OPF normally is in the range of 15–500 μm with cell wall thickness around 3.38 μm and 0.7–1.55 g/cm^3 of density; meanwhile the BET of OPF biochar is around 205.21 m^2/g (Abas and Ani 2014). The physical and chemical properties of OPF normally varied according to their plant species. Table 4.3 shows the proximate and ultimate analysis of oil palm fiber that has been reported by other researchers.

Palm kernel shells (PKS) are the endocarp of oil palm fruit which contains the kernel of the oil palm nut. It is a hard material and can be easily handled in bulk directly from the mill production line. The estimated PKS in 2015 is more than 5 million tonnes/year, and the amount is expected to continuously increase (Wicke

Table 4.4 Lignocellulosic, proximate, and ultimate analyses of PKS (Saka 2005)

Contents	Amounts
Proximate analysis (%)	
Moisture contents	<20
Ash contents	<1.0
Volatile matter	81.03
Fixed carbon	14.87
Lignocellulosic analysis (%)	
Cellulose	20.8
Hemicellulose	22.7
Lignin	50.7
Ultimate analysis (wt %)	
Carbon	53.78
Hydrogen	7.20
Sulfur	0.51
Oxygen	36.30

et al. 2011). PKS have higher heating value compared to other oil palm biomass, and it would be a good source of energy because of its high volatile content. The specific characterization of PKS makes it a good biomass fuel with its low-size variation, easy handling, and limited biological activity due to low moisture content. Nowadays, PKS have being used in cement industries to replace coal for cheaper fuel sources. Table 4.4 shows the characteristics of PKS (Saka 2005).

4.4 Applications of Oil Palm Fiber and Palm Kernel Shell

4.4.1 Activated Carbon in Textile Wastewater Treatment

Activated carbon (AC) is the most widely used adsorbent and has many applications. One of its premier applications is for contaminant removal from water (Merzougui et al. 2011; Ossman et al, 2014) and gases (Lee et al. 2013; Nasri et al. 2014). This is possible due to the highly porous structure of the solid, extended surface area and high degree of surface reactivity. Much of this surface area is contained in mesoporous and microporous structures which resulted in high adsorption capacity of the AC materials. The activation of AC is normally carried out through either physical, chemical, or the combination of physical and chemical activations. In chemical activation, dehydration agents include H_3PO_4, $ZnCl_2$, KOH, and K_2CO_3, each of which has a different effect on the final product AC (Wang et al. 2011). In physical activation, carbon dioxide, steam, and/or air are used for the partial gasification of precursors (Yang and Lua 2003). Researchers are always in search for developing more suitable, efficient, cheap, and easily available types of adsorbents, particularly from the biomass waste materials. Utilizing the palm kernel shell (PKS), for example, would promise a cheaper and more sustainable AC production

Table 4.5 Characteristics of activated carbon from oil palm fiber (OPF) and palm kernel shell (PKS)

Oil palm biomass	Surface area (m^2/g)	Pore size (cm^3/g)	Activation agents	References
Oil palm fiber (OPF)	494	–	Steam	Ibrahim et al. (2017)
	1354	–	KOH/CO_2	Hameed et al. (2008)
	1354	0.778	KOH/CO_2	Tan et al. (2007)
	707.79	2.21 nm	KOH/CO_2	Zaharah Abas and Ani (2016)
Palm kernel shell (PKS)	794	–	Steam	Jia and Lua (2008)
	566.27	–	Steam	Jalani et al. (2016)
	635	–	Steam	Klose and Rincon (2007)
	854.42	0.74	H_3PO_4	Tan et al. (2016)
	1252	0.7233	KOH	Mohammed (2013)
	905	0.569	CO_2	Herawan et al. (2013)
	1170	–	K_2CO_3	Adinata et al. (2007)
	1135	0.80	H_3PO_4/CO_2	Guo and Lua (2003)
	1253.5	2.65	$ZnCl_2$	Zaharah Abas and Ani (2016)

due to the abundance and cheap pricing for PKS. Table 4.5 summarizes the characteristics of AC produced from OPF and PKS.

The uncontrolled discharge of dyes into the environment, notably the water system, is of concern for both toxicological and aesthetical reasons (Pignon et al. 2003). Dyes are used in many industries especially such as paper-making, food technology, hair colorings, and textiles. Commercial dyes consisted of various inorganic compounds which depending on its concentration and long-term usage can lead to toxicity to both fish and other aquatic organisms (Ramakrishna and Viraraghavan 1997). Tan et al. (2007) reported the almost similar capacity of methylene blue removal using AC prepared from OPF and that of commercial AC. The SEM analysis of the OPF-based AC showed that it was porous with well-developed pores with main surface functional groups consist of quinone and aromatic rings (Tan et al. 2007).

4.4.2 *Oil Palm Fiber as Biocomposite in Polymer Industry*

Biocomposites are defined as the materials made of combining natural fiber and petroleum-derived nonbiodegradable polymer or biodegradable polymer where it can be easily disposed or composted without harming the environment (Shinoj et al. 2011; Gurunathan et al. 2015). Due to the increasing environmental sustainability awareness of the public, development of biodegradable and recyclable products such as biocomposites is being given special attention from scientist and policy-makers alike (Faruk et al. 2012). Most of previous studies carried out (Salema et al. 2010; Wong et al. 2010; Then et al. 2013) focused on the reinforcement and modification of the OPF production using either physical (e.g., plasma) or

Table 4.6 Mechanical properties of OPF biocomposite

Reference	Mechanical properties	OPF biocomposite
Hill and Abdul Khalil (2000)		OPF-acetylated
	Length (m)	0.113
	Tensile strength to failure,(MPa)	152
	Tensile modulus (GPa)	5.12
	Strain to failure (%)	13.3
Shinoj et al. (2011)		OPF-polyurethane
	Tensile modulus (MPa)[a]	9.5
	Tensile strength (MPa)[a]	35
	Tensile toughness (MPa)[a]	60
	Flexural modulus (GPa)[b]	2.25
	Flexural strength (MPa)[b]	75
Then et al. (2013)		OPF-poly (butylene succinate)
	Tensile strength (MPa)	15.50 ± 1.04
	Tensile modulus (MPa)	474.20 ± 4.24
	Elongation at break (%)	4.73 ± 0.22
	Flexural modulus (MPa)	1400
	Flexural strength (MPa)	33

[a]At 60% fiber content
[b]At 50% fiber content

chemical treatments (alkaline, enzyme treatment, acetylation, silane). The polymeric matrices used for the fabrication of biocomposites include polyurethane, poly (butylene succinate), epoxy, phenol formaldehyde, polyvinyl chloride and polypropylene, as well as natural rubber. The mechanical properties of OPF biocomposite which has been reinforced with a few polymeric matrices and natural fiber is presented in Table 4.6.

The reinforcing efficiency of natural fiber is closely related to the nature of cellulose and its crystallinity, which is attributable to its cellulose fibril content that provides maximum tensile and flexural strengths, as well as rigidity. Likewise, Then et al. (2013) reported that OPF has better reinforcing ability than that of EFB in PBS biocomposite due to the fact that OPF is softer and more flexible than EFB. This feature gives it higher ability to disperse more evenly and oriented in the PBS matrix, thus resulted in higher tensile strength. In addition, the modification of natural fiber such as OPF by using chemical treatment of cellulosic material normally will alter the physical and chemical structure of the fiber surface. OPF is also less abrasive in handling as well as has minimal impact to potential respiratory problems, i.e., safer to handle and work (Pickering et al. 2016).

4.4.3 Palm Kernel Shell in Concrete Pavement

Palm kernel shell (PKS) has become one of the successful examples for the utilization of waste material in concrete pavement. Today, PKS has been used as material in concrete in order to reduce the negative impacts of concrete toward the environment. Therefore, the development of sustainable lightweight concrete includes replacing up to 25–75% by mass of limestone with PKS (Khankhaje et al. 2016). The concrete produced from this development project is suitable for use in parking lots and traffic roads. There are few factors to consider in applying PKS in concrete pavement such as its physical and mechanical properties along with mechanical strength, durability, functional properties, and structural behavior (Alengaram et al. 2013). The previous lightweight concrete containing PKS showed high water permeability with acceptable compressive strength. According to Chindaprasirt et al. (2008), a good previous concrete has void content between 15% and 25% and strength between 22 and 39 MPa. In order to produce the best previous lightweight concrete, it is imperative to control during the mixture production process and placement method. The utilization of PKS to produce previous lightweight concrete directly helps the construction industry by alleviating the disposal problem and reducing the challenge of scarcity and high cost of cement.

4.5 Biofuel Generation from Oil Palm Fiber and Palm Kernel Shell

The oil production from large oil fields is declining at the rate of 4–5% annually, but the global production of oil and gas is approaching its maximum. Moreover, the production and consumption of fossil fuels have become a concern with respect to the global climate change. Presently, biofuels production has increased around the world due to the rise in world crude oil prices and also in efforts to reduce the GHG emission (Kurnia et al. 2016). Biofuels can be referred to as liquid, gas, and solid fuels for the power generation sector which is mainly produced from renewable sources such as biomass.

4.5.1 Biomass Gasification

Biomass gasification is a thermochemical process to obtain gaseous fuel product with several potential applications. Oil palm biomass such as PKS, OPF, and EFB are the potential feedstocks for the gasification process in order to produce hydrogen gas (Shuit et al. 2009). The thermochemical partial oxidation process converts the biomass solid residue into usable gaseous fuel in the presence of gasifying agents, such as air, steam, oxygen, carbon dioxide, or a mixture of these agents. It is a chemical process that generate synthesis gas which could be used either as a gaseous fuel for power plants and transportation sectors or as a feedstock for chemical industries.

Clean gaseous fuel could be used for internal combustion engines and gas turbines for power generation. In addition, there are many factors that lead to the quality of synthesis gas such as the feedstock material, the presence of catalyst, design of the reactor, gasifying agent, as well as the operational conditions of the reactor. Ideally, the oil palm fiber and palm kernel shell in the form of char (rather than its raw dried form) are to be used for the gasification process in order to produce the gases that is free of water, tar, as well as the other corroding components (Ani 2012). There are various conventional biomass gasification technologies available which includes fixed bed (updraft, downdraft, and crossdraft), fluidized bed, circulating fluidized bed, and entrained flow reactors. The latest development for biomass gasification approach is via the use of supercritical water technology and the biomass integrated gasification combined cycle which is initially been designed and operated for coal gasification (Farzad et al. 2016). Indeed, the gasification of biomass and subsequent conversions have led to environmental sustainability; the development of regional economic, social, and agricultural; as well as the reduction in GHG emissions.

4.5.2 Briquetting and Densification

Briquetting or densification is a mechanical process of compacting biomass into a uniform solid fuel called briquettes. It is one of the most desired techniques to improve the storage and transportation of the fuel in order to prevent dumped areas adjacent to palm mills and becoming another waste product (Bhattacharya et al. 2002). The densification process also is indirectly responsible in the reduction of dust formation and improves the combustion properties of biomass materials such as calorific value, moisture content, and burning rate (Faizal et al. 2010; 2015). This approach is notable in expanding the use and marketability of palm biomass fuel either for domestic or export market (Nasrin et al. 2011).

In general, briquetting of biomass can be carried out using several techniques with or without the presence of binding agents. Commercially, there are two technologies involved, namely, screw extrusion and piston press technologies (Hasan et al. 2016). Basically, the densification products can be categorized into three types based on the diameter, namely, pellet, briquette, and bale. However, the densification of palm oil biomass is only closely related to pellet and briquette, and this is due to the necessary chopping and milling process before the densification process. Densification technique has been applied for palm biomass around 16 years ago, specifically briquetting process for PKS and OPF (Husain et al. 2002; Chin et al. 2000). Sing and Aris (2012) reported that the compressive strength of 100% OPF briquette is about 2.5 times higher than that of 100% PKS briquette, which can be attributed to the fibrous structure of the OPF which is responsible to hold the whole briquette more firmly. A mixture of 60% PKS and 40% OPF, with paper as its binding agent, was determined to be the optimum ratio for a viable solid fuel (Sing and Aris 2013). The presence of high shell content in the mixture of OPF and PKS briquettes also increased the calorific value, specific density, and quality of the briquette as well (Nasrin et al. 2011).

4.5.3 Combustion Fuel for Steam Generation

Malaysia, Indonesia, and Thailand have a huge potential in palm waste power generation based on the huge amount of biomass produced yearly. This represents a projected palm oil extract to dry biomass waste produced ratio of 1: 4 kg after extraction process (Yusoff 2006). All palm oil mills generally meet most of their electricity and process steam requirements by burning some of the oil palm wastes, with energy for start-up generally being provided by backup diesel. OPF and PKS mixtures are usually used in palm oil industries as fuel for their steam generation in its electrical power and thermal heating. The chemical energy produced during combustion in the boiler is converted to heat energy. For each kg of palm oil produced, about 0.075–0.1 kWh of electricity is consumed, and its steam demand around 2.5 kg (Yusoff 2006). The electricity cogenerated in Malaysian palm oil mills, therefore, only contributed around 1–1.5 billion kWh or less than 2% of the total electricity generation in Malaysia, i.e., >82 billion kWh (Sulaiman et al. 2011).

4.6 Future Trends

Malaysian palm oil industries are improving the technologies in mill processing of the palm oil by developing its own local technology. The authority should act quickly to promote investors and private sectors to develop such improved local technologies from the current technology. Palm oil mill plant performances can be improved from years of operational experiences. These activities also help the local people to be creative and innovative and can improve the current living standards for future generations. New compact mill with smaller processing capacity for countries with small holding plantations that consolidate together utilizing the main grid power should be applied and shared in the region. Technologies are feasible if they are fabricated, operated, and managed locally. Further improvements would be achieved as time goes by, and this is how the developed countries develop themselves to be self-sustained in encountering problems. The potential to obtain various products from oil palm biomass such as bio-oil (through fast pyrolysis techniques), pyroligneous acid (from slow pyrolysis process), and biochar has continued to form the core motivation in the planning of oil palm biomass management strategies.

4.7 Conclusions

Oil palm cultivation holds an important asset for its food sources and other applications from its oil and biomass. The biomass products have a great potential for a sustainable and distributed energy community sectors. However, the choice to use biomass for power, fuel, and biomaterials depends on a variety of factors, such as availability, public policy, cost of biomass, capital cost of processing equipments, infrastructure facilities, and markets for alternative energy and materials. The development of advanced conversion technologies suitable for different types of biomass is likely to make biomass both energy and materials products competitive with petroleum-based products. New application of microwave-assisted processing of bio-oil, biochar, activated carbon, and fine biochemicals shows a great potential in a sustainable integrated processing plant. The integrated processing plant of the palm fruits using excess energy for modular processing could be used in the future for the palm mills operation. In addition, these challenges and solutions can be addressed and advanced by progressive country policies, technological progress, and social awareness of environmental problems.

References

Abas FZ, Ani FN (2014) Comparing characteristics of oil palm biochar using conventional and microwave heating. Jurnal Teknologi (Sci Eng) 68(3):33–37

Abnisa F, Arami-niya A, Wan Daud WMA, Sahu JA (2013) Characterization of bio-oil and bio-char from pyrolysis of palm oil wastes. Bioenergy Res 6(2):830–840

Adinata D, Daud WM, Aroua MK (2007) Preparation and characterization of activated carbon from palm shell by chemical activation with K_2CO_3. Bioresour Technol 98(1):145–149

Agensi Inovasi Malaysia (2012) National biomass strategy delivery unit (1MBAS): promote, coordinate and facilitate the business and investment opportunities in Malaysia's biomass based industries. Retrieved Jan 15, 2017 from http://www.nbs2020.gov.my/1mbas-2012

Alengaram UJ, Muhit BA, Jumaat MZ (2013) Utilization of oil palm kernel shell as lightweight aggregate in concrete – a review. Constr Build Mater 38:161–172

Ani FN (2012) Sustainability and recycling through thermal conversion of bioresources. Professorial inaugural lecture series. Penerbit UTM Press, pp 1–42. ISBN 978-983-52-0861-4

Aziz MA, Uemura Y, Sabil KM (2011) Characterization of oil palm biomass as feed for torrefaction process. National Postgraduate Conference (NPC). 19–20 September. IEEE, Perak, pp 1–6

Bhattacharya SC, Leon MA, Rahman MM (2002) A study on improved biomass briquetting. Energy Sustain Dev 6(2):67–71

Chen WH, Lin BJ (2016) Characteristics of products from the pyrolysis of oil palm fiber and its pellets in nitrogen and carbon dioxide atmospheres. Energy 94:569–578

Chindaprasirt P, Hatanaka S, Chareerat T, Mishima N, Yuasa Y (2008) Cement paste characteristics and porous concrete properties. Constr Build Mater 22(5):894–901

Choi G-G, Seung-Jin O, Lee S-J, Kim J-S (2015a) Production of bio-based phenolic resin and activated carbon from bio-oil and biochar derived from fast pyrolysis of palm kernel shells. Bioresour Technol 178:99–107

Choi G-G, Seung-Jin O, Lee S-J, Kim J-S (2015b) Production of bio-based phenolic resin and activated carbon from bio-oil and biochar derived from fast pyrolysis of palm kernel shells. Bioresour Technol 178:99–107

Cookwithkathy (2013) Retrieved on Jan 23, 2017 from https://cookwithkathy.wordpress.com/2013/08/02/what-is-the-difference-between-palm-oil-and-coconut-oil/

Faizal HM, Latiff ZA, Mazlan AW, Darus AN (2010) Physical and combustion characteristics of biomass residues from palm oil mills. New aspects of fluid mechanics, heat transfer and environment. WSEAS Press, Taiwan

Faizal HM, Latiff ZA, Mohd Iskandar MA (2015) Characteristics of binderless palm biomass briquettes with various particle sizes. Jurnal Teknologi 77(8):1–5

Faruk O, Bledzki AK, Fink HP, Sain M (2012) Biocomposites reinforced with natural fibers: 2000–2010. Prog Polym Sci 37(11):1552–1596

Farzad S, Mandegari MA, Görgens JF (2016) A critical review on biomass gasification, co-gasification, and their environmental assessments. Biofuel Res J 3(4):483–495
GGS (2013) http://www.ggs.my/index.php/palm-biomass. Accessed 23 Oct 2013
Guo J, Lua AC (2003) Textural and chemical properties of adsorbent prepared from palm shell by phosphoric acid activation. Mater Chem Phys 80:114–119
Gurunathan T, Mohanty S, Nayak SK (2015) A review of the recent developments in biocomposites based on natural fibres and their application perspectives. Compos A: Appl Sci Manuf 77:1–25
Hameed BH, Tan IAW, Ahmad AL (2008) Optimization of basic dye removal by oil palm fibre-based activated carbon using response surface methodology. J Hazard Mater 158:324–332
Hasan MF, Rahman MRA, Latiff ZA (2016) Review on densification of palm residues as a technique for biomass energy utilization. Jurnal Teknologi 78(9–2):9–18
Herawan SG, Hadi MS, Ayob MR, Putra A (2013) Characterization of activated carbons from oil-palm shell by CO_2 activation with no holding carbonization temperature. Sci World J 2013:1–6
Hill CAS, Abdul Khalil HPS (2000) Effect of fiber treatments on mechanical properties of coir or oil palm fiber reinforced polyester composites. J Appl Polym Sci 78(9):1685–1697
Husain Z, Zainac Z, Abdullah Z (2002) Briquetting of palm fibre and shell from the processing of palm nuts to palm oil. Biomass Bioenergy 22(6):505–509
Ibrahim I, Hassan MA, Aziz SA, Shirai Y, Andou Y, Othman MR, Ali AAM, Zakaria MR (2017) Reduction of residual pollutants from biologically treated palm oil mill effluent final discharge by steam activated bioadsorbent from oil palm biomass. J Clean Prod 141:122–127
Jalani NF, Aziz AA, Wahab NA, Hassan WH, Zainal NH (2016) Application of palm kernel shell activated carbon for the removal of pollutant and color in palm oil mill effluent treatment. J Earth Environ Health Sci 2:15–20
Jia Q, Lua CA (2008) Effects of pyrolysis conditions on the physical characteristics of oil-palm-shell activated carbons used in aqueous phase phenol adsorption. J Anal Appl Pyrolysis 83:175–179
Khalil HA, Hassan A, Zaidon A, Jawaid M, Paridah MT (2012) Oil palm biomass fibres and recent advancement in oil palm biomass fibres based hybrid biocomposites. In: Composites and their applications. InTech, Croatia, pp 187–220
Khankhaje E, Salim MR, Mirza J, Hussin MW, Rafieizonooz M (2016) Properties of sustainable lightweight pervious concrete containing oil palm kernel shell as coarse aggregate. Constr Build Mater 126:1054–1065
Klose W, Rincon S (2007) Adsorption and reaction of NO on activated carbon in the presence of oxygen and water vapour. Fuel 86:203–209
Kurnia JC, Jangam SV, Akhtar S, Sasmito AP, Mujumdar AS (2016) Advances in biofuel production from oil palm and palm oil processing wastes: a review. Biofuel Res J 3(1):332–346
Lee CS, Ong YL, Aroua MK, Daud WMAW (2013) Impregnation of palm shell-based activated carbon with sterically hindered amines for CO_2 adsorption. Chem Eng J 219:558–564
Merzougui Z, Azoudj Y, Bouchemel N, Addoun F (2011) Effect of activation method on the pore structure of activated carbon from date pits application to the treatment of water. Desalin Water Treat 29:236–240
Mohammed RR (2013) Decolorisation of biologically treated palm oil mill effluent (POME) using adsorption technique. Int Refereed J Eng Sci 2(10):1–11
Nasri NS, Hamza UD, Ismail SN, Ahmed MM, Mohsin R (2014) Assessment of porous carbons derived from sustainable palm solid waste for carbon dioxide capture. J Clean Prod 71:148–157
Nasrin AB, Choo YM, Lim WS, Joseph L, Michael S, Rohaya MH, Astimar AA, Loh SK (2011) Briquetting of empty fruit bunch fibre and palm shell as a renewable energy fuel. J Eng Appl Sci 6(6):446–451
Ossman ME, Abdel Fatah M, Taha NA (2014) Fe(III) removal by activated carbon produced from Egyptian rice straw by chemical activation. Desalin Water Treat 52:3159–3168
Pickering KL, Efendy MA, Le TM (2016) A review of recent developments in natural fibre composites and their mechanical performance. Compos A: Appl Sci Manuf 83:98–112

Pignon HM, Brasquet CF, Cloirec PL (2003) Adsorption of dyes onto activated carbon cloths: approach of adsorption mechanisms and coupling of ACC with ultrafiltration to treat coloured wastewaters. Sep Purif Technol 31:3–11

Piskorz J, Radlein D, Scott DS, Czernik S (1988) Liquid products from fast pyrolysis of wood and cellulose. Elsevier Appl Sci: 557–571

Ramakrishna KR, Viraraghavan T (1997) Dye removal using low cost adsorbents. Water Sci Technol 36(2–3):189–196

Sabil KM, Aziz MA, Lal B, Uemura Y (2013) Effects of torrefaction on the physiochemical properties of oil palm empty fruit bunches, mesocarp fiber and kernel shell. Biomass Bioenergy 56:351–360

Saka S (2005) Whole efficient utilization of oil palm to value-added products. In: Proceedings of JSPS-VCC Natural Resources & Energy Environment Seminar, Kyoto, Japan

Salema A, Hassan A, Bakar AA (2010) Oil-palm fiber as natural reinforcement for polymer composites. Plast Res Online

Shinoj S, Visvanathan R, Panigrahi S, Kochubabu M (2011) Oil palm fiber (OPF) and its composites: a review. Ind Crop Prod 33(1):7–22

Shuit SH, Tan KT, Lee KT, Kamaruddin AH (2009) Oil palm biomass as a sustainable energy source: a Malaysian case study. Energy 34(9):1225–1235

Sing CY, Aris MS (2012) An experimental investigation on the handling and storage properties of biomass fuel briquettes made from oil palm mill residues. J Appl Sci 12(24):2621–2625

Sing CY, Aris MS (2013) A study of biomass fuel briquettes from oil palm mill residues. Asian J Sci Res 6(3):537–545

Sulaiman F, Abdullah N, Gerhauser H, Shariff A (2011) An outlook of Malaysian energy, oil palm industry and its utilization of wastes as useful resources. Biomass Bioenergy 35(9):3775–3786

Tan IAW, Hameed BH, Ahmad AL (2007) Equilibrium and kinetic studies on basic dye adsorption by oil palm fibre activated carbon. Chem Eng J 127:111–119

Tan IAW, Chan JC, Hameed BH, Lim LLP (2016) Adsorption behavior of cadmium ions onto phosphoric acid-impregnated microwave-induced mesoporous activated carbon. J Water Process Eng 14:60–70

Then YY, Ibrahim NA, Zainuddin N, Ariffin H, Wan Yunus WMZ (2013) Oil palm mesocarp fiber as new lignocellulosic material for fabrication of polymer/fiber biocomposites. Int J Polym Sci 2013

Wang X, Liang X, Wang Y, Wang X, Liu M, Yin D, Xia S, Zhao J, Zhang Y (2011) Adsorption of Copper (II) onto activated carbons from sewage sludge by microwave-induced phosphoric acid and zinc chloride activation. Desalination 278:231–237

Wicke B, Sikkema R, Dornburg V, Faaij APC (2011) Exploring land use changes and the role of palm oil production in Indonesia and Malaysia. Land Use Policy 28(1):193–206

Wong KJ, Nirmal U, Lim BK (2010) Impact behavior of short and continuous fiber reinforced polyester composites. J Reinf Plast Compos 29(23):3463–3474

Yang T, Lua AC (2003) Characteristics of activated carbons prepared from pistachio-nut shells by physical activation. J Colloid Interface Sci 267:408–417

Yusoff S (2006) Renewable energy from palm oil-innovation on effective utilization of waste. J Clean Prod 14(1):87–93

Zaharah Abas F, Ani FN (2016) Characteristic of oil palm activated carbon produced from microwave and conventional heating. In Appl Mech Mater 819:606–611

Chapter 5
Optimization of Ferulic Acid Production from Oil Palm Frond Bagasse

Zulsyazwan Ahmad Khushairi, Hafizuddin Wan Yussof, and Norazwina Zainol

Abstract Ferulic acid (FA) is an organic acid that possesses multiple physiological properties including anti-oxidant, anti-microbial, anti-inflammatory, anti-thrombosis and anti-cancer activities. The applications of FA include being the source for vanillin and preservative production, thin film for food packaging, food supplement and skin care products. Oil palm frond (OPF) is the leaf and the branch part of the oil palm tree. In Malaysia, OPF is found in abundance as it is one of the by-products of the palm oil industry. The oil palm frond bagasse (OPFB) is obtained after some treatments that include pressing and drying of OPF, of which the resulting fibre is used for subsequent processes. Choice of using enzymatic hydrolysis to produce FA is more attractive compared to conventional chemical hydrolysis as enzymatic hydrolysis mainly focuses on utilizing the reaction caused by feruloyl esterase (FAE) to release FA from polysaccharide.

5.1 Soil Culture Role in Ferulic Acid Production

Various mesophilic soil bacteria and fungi have the ability to produce FAEs. These include *A. niger*, *Streptomyces* sp. and *Penicillium* sp. The FAEs purified from *A. niger* have different physicochemical characteristics and catalytic properties against cinnamoyl model substrates. The methyl derivatives of hydroxycinnamic esters and soluble feruloylated oligosaccharides derived from plant cell wall are the notable differences of the *A. niger* (Ralph et al. 1994). These enzymes are exocellular and possess such inducible corresponding expression. *A. niger* I-1472 showed the tendency to produce numerous polysaccharide-degrading enzymes including esterases that released ferulic acid from natural feruloylated oligosaccharides on sugar beet pulp and maize bran (Szwajgier et al. 2005).

Z.A. Khushairi • H.W. Yussof (✉) • N. Zainol
Faculty of Chemical and Natural Resources Engineering, Universiti Malaysia Pahang, Kuantan, Pahang, Malaysia
e-mail: hafizuddin@ump.edu.my

Z.A. Zakaria (ed.), *Sustainable Technologies for the Management of Agricultural Wastes*, Applied Environmental Science and Engineering for a Sustainable Future, https://doi.org/10.1007/978-981-10-5062-6_5

The aspergilli, a common fungus family that can be found in soil, was determined to be one of the feruloyl esterase (FAE) enzyme producer (Gopalan et al. 2015). This approach is mainly focused on the ability of mixed cultures to produce and enhance the enzyme activity, hence allowing the enzymatic hydrolysis to take place. One of the reasons behind this approach is to apply enzymatic hydrolysis using mixed cultures propagated at a much lower cost as well as higher purity ferulic acid compared to that obtained using the chemical hydrolysis.

5.2 Factors Affecting Ferulic Acid Production via Soil Culture Treatment

Ferulic acid production via enzymatic release has its efficiency related directly to the type of substrate used. Agricultural-based materials are an excellent choice for feedstock based on its high ferulic acid production. Other factors or parameters that may influence the efficiency of the process include pre-treatment application. The parameters during ferulic acid production itself will also affect the outcome of the process. Factors such as temperature, reaction time and agitation are some examples of parameters that will affect ferulic acid production in enzymatic hydrolysis (Gopalan et al. 2015).

5.2.1 Temperature

Temperature is the most common parameter for any hydrolysis process. The effect of temperature to FAEs and other enzymes during hydrolysis has been extensively studied by a number of researchers (He et al. 2015; Tsang et al. 2006; Xun et al. 2012). In terms of FAEs, in order to achieve high enzyme activity, the perfect temperature is required. He et al. (2015) reported that when the incubation temperature was varied between 30 and 70 °C, the maximum activity of FAEs was determined at 50 °C. Similar to most chemical reactions, the rate of an enzyme reaction increases as the temperature increases. However, at high temperature, enzymes are exposed to the risk of being unstable, causing it to be denatured. There is no fixed temperature value that can maximize enzyme activity (Bisswanger 2014). The range of temperature used in enzymatic hydrolysis is crucial for the process to be successful.

5.2.2 pH

pH also plays an important role to enzyme activity where certain enzyme may only be active at the required pH value. An extremely high or low pH value will almost certainly result in a complete loss of enzyme activity for most enzymes. The pH value also plays a part in the stability of enzymes. For each enzyme activity, there is a region of pH optimal stability. Studies showed that the optimum pH value for FAE activity is around pH 6.0 (Abraham et al. 2014; He et al. 2015; Huang et al. 2003). These studies also claimed that pH value does contribute a high impact to FAE activity. The FAE has shown to be consistently active at 60% of the optimal activity between pH 5 and 7.5 (Kühnel et al. 2012).

5.2.3 *Agitation Rate*

The agitation rate during enzymatic hydrolysis is essential as it will affect the volume of dissolved oxygen in the cultivation medium. Excessive agitation rate could risk the process to the condition of hydrodynamic shear stresses. This could affect the fungal mycelia and pellets (Rodríguez Porcel et al. 2005). Damaged fungal mycelia and pellets would lead to cell devastation, lowering the production of enzyme. The cell growth, redox balance, mitochondria function and generation of energy during fermentation process were found to be improved when agitation rate is applied (Silva et al. 2010). This improvement to the fermentation process will directly improve the FAE activity in the enzymatic hydrolysis process.

5.2.4 *Inoculum and Substrate Volume Percentage*

The balance between both inoculum and substrate volumes also played a critical part in influencing the enzymatic hydrolysis process. One of the drawbacks of having high substrate percentage includes substrate inhibition, causing the reduction in enzyme formation (Shahzadi et al. 2014). Different substrate volume will affect the FAE for FA production. However, low substrate volume could lead to insufficient source of energy for the inoculum. The balance between inoculum and substrate volumes is essential in enzymatic hydrolysis process. One of the efficient ways to improve this factor is to fix the inoculum volume while improving the substrate volume until the reaction rate is at maximum. Almost all of the commonly used inoculum and substrate volumes for enzymes have been determined through Michaelis constants. However, this is not applicable for all types of enzyme (Martinek 1969).

5.2.5 *Reaction Time*

FA production is also influenced by the time of the reaction. A study showed that the highest amount of FA was discovered at 45 °C, around 30 min after the start of a mashing process (McMurrough et al. 1996). The FA levels remained the same after this period. This situation could be attributed to protein denaturation resulting in the loss of catalytic activity. In another study, the enzyme production was found to be directly related to the growth of microorganism (Mohanasrinivasan et al. 2014). The growth would reach its limit at the point of nutrient deficiency. On increasing the incubation beyond 72 h, the study discovered that the enzyme activity decreased due to denaturation. This further improves the importance of reaction time in enzymatic hydrolysis process.

5.3 Screening of Factors Affecting Ferulic Acid Production

The extent of different factors towards the production of ferulic acid was determined by Khushairi et al. (2016) using the two-level factorial design analysis. A 2^5 factorial design at half fraction was selected as the experimental design. Five factors were analysed for their effect on the production of ferulic acid. The five factors were temperature, pH, agitation rate, reaction time and inoculum volume percentage.

5.3.1 Contribution of Factors

The contribution of each factor is shown in Table 5.1. The pH value stood as the factor with the highest percentage contribution to the process at 23.76%, followed by temperature at 23.34%. As for agitation rate, time and inoculum percentage, these three factors showed low values of percentage contribution at 5.77%, 1.20% and 3.39%, respectively. Ferulic acid production process is crucially affected by FAE. Similar to other enzymes, pH and temperature play a critical role in inducing FAE relative activity. The optimum value of these factors differed from one FAE to another as FAE itself has a wide range of substrates and sources (Esteban-Torres et al. 2013). A study on the temperature and pH profiles of multiple FAEs (Li et al. 2011) discovered that three FAEs in *Cellulosilyticum ruminicola* have an optimum pH range from 6 to 9 and temperature range from 35 to 40 °C. This shows that both pH value and temperature affect ferulic acid production.

Table 5.1 Percentage contribution of each main factor to ferulic acid production

Factor	Percentage contribution (%)
Temperature	23.34
pH value	23.76
Agitation rate	5.77
Reaction time	1.20
Inoculum percentage	3.39

5.3.2 Interaction Between Factors

From the factorial analysis, there were a total of four interactions discovered throughout the process. These four interactions are; (1) temperature and pH, (2) temperature and agitation rate, (3) pH and percentage inoculum, and (4) agitation rate and time.

5.3.2.1 Temperature and pH Value

As depicted in Fig. 5.1, at 26 °C, the yield between pH 9 and pH 5 was far apart from each other as pH 9 showed to be the one with the high yield. At 40 °C, pH 9 was still the one with a higher yield, but the yield difference between the two pH values has decreased. This shows that there exists an interaction between temperature and pH. Temperature gave high impact to the release of FAE (Ou et al. 2011). With the presence of FAE in the sample, the cross-link that inhibited ferulic acid production can be broken down, notably when FAE is at its most active state, i.e. optimal pH and temperature conditions (Wang et al. 2004). The interaction of these two factors contributed to FAE activity and hence the ferulic acid production.

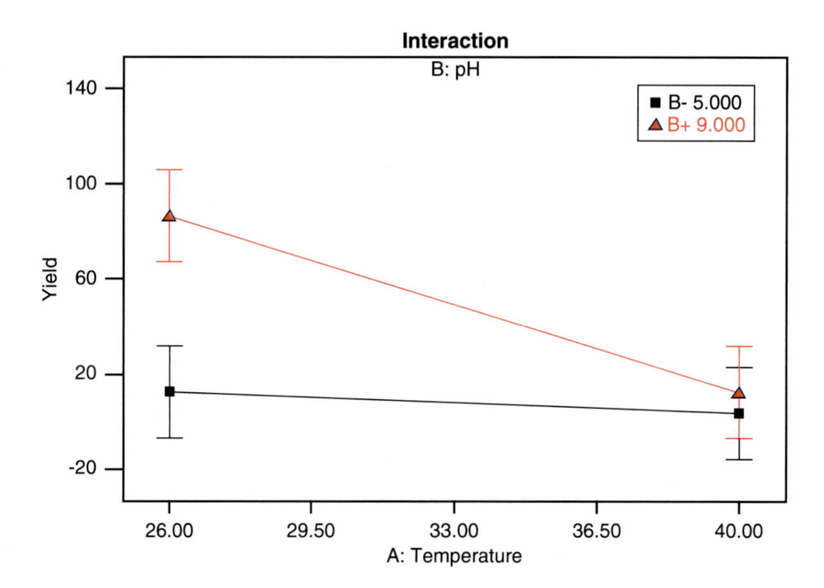

Fig. 5.1 Interaction between temperature (A) and pH (B)

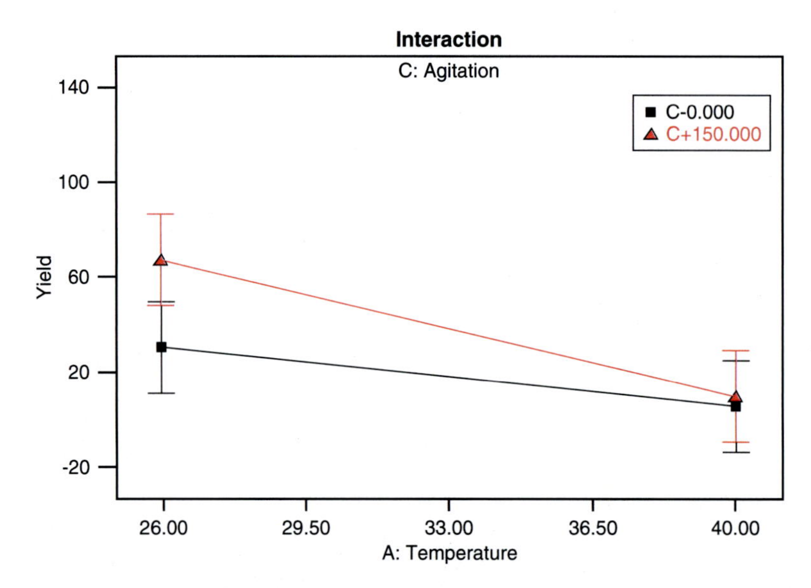

Fig. 5.2 Interaction between temperature (A) and agitation speed (C)

5.3.2.2 Temperature and Agitation Rate

From Fig. 5.2, at temperature 26 °C, high agitation speed resulted in higher yield compared to no agitation speed. However, at 40 °C, the yield for both agitation speeds is almost identical. This interaction trending is almost identical with interaction between temperature and pH value but at a smaller scale. From the data obtained, it is clear that agitation plays a small role in the production of ferulic acid. Faulds et al. (1997) stated that *A. niger* requires 25 °C and 150 rpm of agitation for fungal growth. This suggest that at a certain combination of agitation speed and temperature, FAE will be released, contributing to the ferulic acid production.

5.3.2.3 pH and Inoculum Percentage

Figure 5.3 shows the interaction between pH and percentage inoculum. At pH 5, 10% inoculum was found to be producing a higher yield of FA compared to 2% inoculum. At pH 9, 2% inoculum showed to be yielding at a higher value compared to 10% inoculum. At around pH 5.8, both 2% inoculum and 10% inoculum produced the same yield amount of ferulic acid. The pH value has shown to be inducing the FAE activity (Li et al. 2011). At the optimal pH value, FAE activity can be utilized to its best performance. As for inoculum, the lower volume of inoculum proves to be more efficient compared to the higher volume. A study on FA production from *A. niger* showed that on common practice, inoculum at low volume is much preferred (Ou et al. 2011).

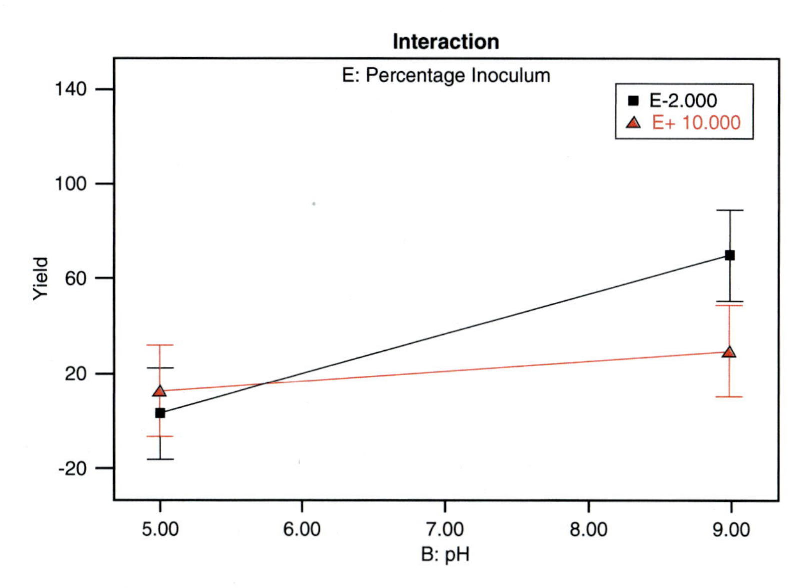

Fig. 5.3 Interaction between pH (B) and percentage inoculum (E)

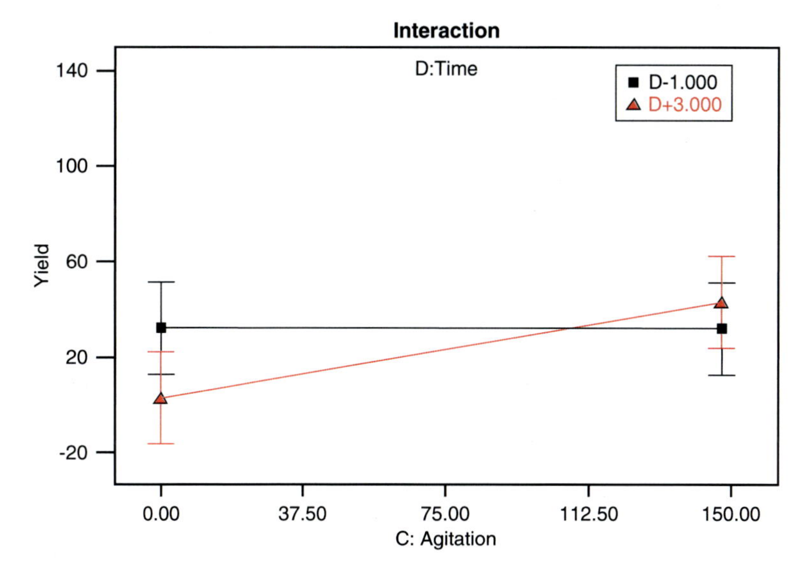

Fig. 5.4 Interaction between agitation (C) and time (D)

5.3.2.4 Agitation Rate and Time

The interaction between agitation (C) and time (D) is as shown in Fig. 5.4. When no agitation was applied, the yield for 1-day is shown to be higher than 3-day. However, at 50 rpm, the yield for 3-day has a higher value than the yield for 1-day. Agitation speed has proved to be one of the most significant contributing factors for ferulic acid production. However, optimal agitation speed for various raw materials will differ (Gopalan et al. 2015). The same concept applies to the interaction time. At 150 rpm, the difference in ferulic acid yield for both 3-day and 1-day duration is too small to make any significance. In this case, a 1-day duration is much more preferred as the process will consume less time.

5.4 Optimization of Ferulic Acid Production

Response surface methodology (RSM) based on the central composite design (CCD) was used for experimental design for optimization process. Two of the screened factors from the factor analysis process were used as the independent factors. With a fit of a second-order (quadratic) model for the full factorial CCD, the design consisted of 13 sets of experiments including 8 samples and 5 centre point. Five levels of variation of numeric factor were used in the experiments, respectively. The five levels consisted of plus and minus alpha (axial point), plus and minus 1 (factorial points) and the centre point. The two factors selected for the optimization process were temperature and agitation rate. Although from the screening process pH value showed as the most significant contributing factor, pH was not selected for the optimization process as it may affect other reaction to occur. In this case, the increase in pH value may expose the process to alkaline hydrolysis. In order to avoid such event, the pH value was fixed at pH 9. Similarly, reaction time was also fixed at 1-day and inoculum percentage at 2%. The value of each run with their respective responses is shown in Table 5.2.

5.4.1 Statistical Modelling and ANOVA

The selected variables from the factor analysis process were temperature (factor A) and pH value (factor B). These variables were studied, and the response was designated as percentage of final lignin content. With the aid of Design-Expert software, 13 sets of experiments were formed, and a quadratic model was proposed. Similar to the factor analysis process, Design-Expert software 7 was used to develop the experimental plan and optimize the regression equation. Table 5.2 presented the full design of CCD for optimization process along with its variables and the yield.

Table 5.3 shows the analysis of variance table (ANOVA) for optimization process. The ANOVA for optimization experimental design showed that the R^2 was 0.8791. The model F value of 10.18 implies the model is significant. There is only a 0.41% chance that the model could occur due to noise.

Table 5.2 Yield of ferulic acid for each run

No	Temperature (°C)	Agitation speed (rpm)	Yield (mg FA/kg OPFB)
1	−1	−1	209.7175
2	+1	−1	183.6000
3	0	+1	168.9531
4	+1	+1	165.7292
5	−α	0	145.8828
6	+α	0	149.6126
7	0	−α	131.7538
8	0	+α	138.5850
9	0	0	203.1590
10	0	0	200.0238
11	0	0	205.7242
12	0	0	201.6246
13	0	0	202.4222

Table 5.3 Analysis of variance table (ANOVA) for factor optimization process

Source	Sum of squares	df	Mean square	F value	Prob > F	
Model	8173.8998	5	1634.7800	11.0182	0.0033	Significant
A-temperature	15.7398	1	15.7398	0.1061	0.7542	
B-agitation rate	113.0636	1	113.0636	0.7620	0.4116	
AB	54.42835	1	54.4283	0.3668	0.5638	
A^2	4017.5154	1	4017.5154	27.0776	0.0012	
B^2	6152.2585	1	6152.2585	41.4655	0.0004	
Residual	1038.5933	7	148.3705			
Lack of fit	854.6012	3	284.8671	6.1930	0.0552	Not significant
Pure error	183.9920	4	45.9980			
Cor total	9212.4931	12				
Std. deviation	12.18					
Mean	176.32					
R^2	0.8873					

5.4.2 *Response Surface Plots*

Table 5.4 shows the coefficient regression for linear equation in optimization process. This table was propagated from the final equation in terms of coded factor in the ANOVA. The response surfaces were plotted by using Eq. 5.1 in the next step. This response surfaces mapped the relationship between the response and variables according to the full model.

Table 5.4 Coefficient regression for linear equation regression in optimization process

Factor	Coefficient estimate
a_0	203.67
a_1	−1.15
a_2	−3.07
a_1a_2	3.69
a_1a_1	−13.24
a_2a_2	−16.39

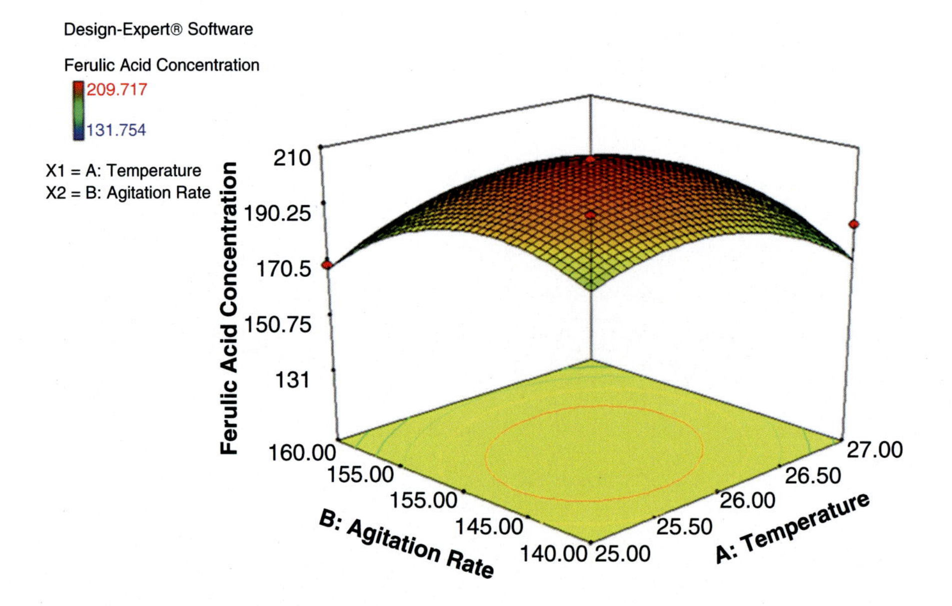

Fig. 5.5 3D surface view of the model for optimization

From Table 5.4, the final equation in terms of coded factors for optimization process:

$$Y = 203.67 - 1.15A - 3.07B + 3.69AB - 13.24A^2 - 16.39B^2 \quad (5.1)$$

The response surface plots unveiled an optimum point in the ferulic acid production process. The response surface plots showed plateau shape of contour. This proved that there exists a point where the yield is the highest for the process. The existence of the point with the highest value proved that optimum point was located in the studied range. Figure 5.1 shows the response surface plots of the process with factor B as the x-axis, response as the y-axis and factor A as the z-axis. Since there were two factors involved in the optimization study, there was only one response surface plot figure generated from Design-Expert. From Fig. 5.5, it is clearly shown that there is a peak in the response surface plot. This proved that the optimum point exists in the studied range. The trending of the response surface plot indicates that

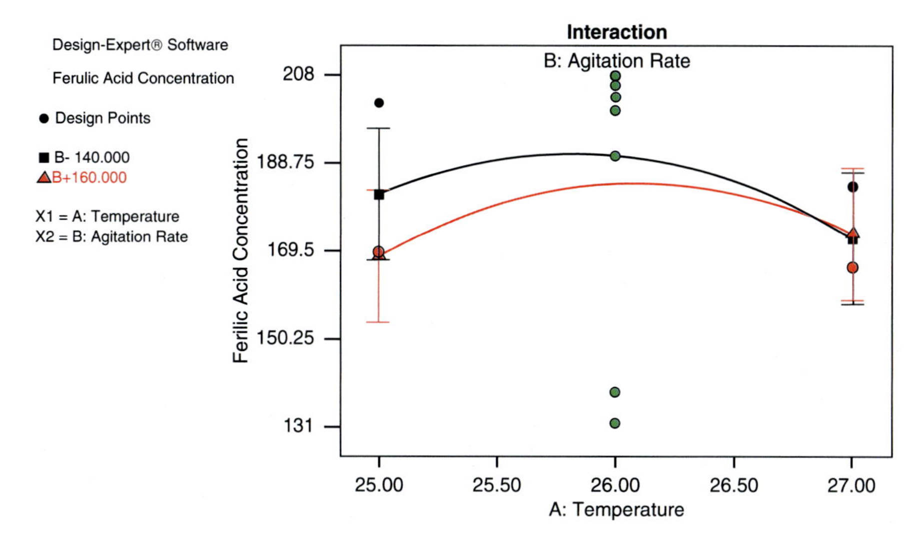

Fig. 5.6 Interaction between temperature (factor A) and agitation rate (factor B)

the optimum point is located towards the centre point of the range studied for both factor A and factor B.

5.4.3 Interaction Between Factors

During the optimization process, both of the factors involved may affect one and the other. This phenomenon may improve the yield of the process or give the exact opposite effect. Since there were only two factors involved, there was only one factor interaction occurred, i.e. between factor A and factor B (Fig. 5.6). At 25 °C, agitation rate at 140 rpm is shown to be yielding a higher value compared to the agitation rate at 160 rpm. However, the opposite trend was observed at 27 °C where the use of 160 rpm resulted in a slightly higher yield compared to 140 rpm.

This factor interaction trend directly indicates the significant role of agitation speed towards the improvement of FA yield when operated at higher temperature. In view of this, 26 °C can be said as the optimum temperature to obtain maximum yield of FA. At this temperature, the inoculum was able to release FAE at its optimum level, hence allowing the breakdown of the polysaccharide structure.

5.4.4 Data Validation

Outcome from the validation process for this model is shown in Table 5.5. This value will be used in other objectives such as the reaction mechanism, kinetic modelling and scale-up process. The theoretical value was obtained from Design-Expert

Table 5.5 Validation process result and error value

Replicates	Temperature (°C)	Agitation (rpm)	Theoretical value (mg FA/kg OPFB)	Experimental value (mg FA/kg OPFB)	Percentage of error (%)
1	26	150	205.724	192.9798	6.1948
2	26	150	205.724	192.1649	6.5909
3	26	150	205.724	180.1041	12.4535
				Average error	8.4131

under point-prediction settings. An average error of 8.4131% showed that the experiment is repeatable with a small error. Therefore, it can be concluded that optimum condition for the production of ferulic acid from OPFB using mixed cultures is as follows: temperature at 26 °C, agitation speed at 150 rpm, pH value at pH 9, time at 1-day and inoculum percentage at 2%.

5.5 Conclusion

The screening process was successful to determine the factors that played a major role in ferulic acid production. Five factors were studied (temperature, pH value, agitation speed, time and inoculum percentage), and their percentage contribution were 23.34%, 23.76%, 5.77%, 1.20% and 3.39%, respectively. There were four interactions discovered throughout the study which were the interaction between temperature and pH value, temperature and agitation speed, pH value and percentage inoculum, and agitation speed and time. Only temperature and agitation speed were studied, while the other factors were fixed at certain values. Validation experiment showed an average error of 8.4131%, hence proving that the experiment is repeatable. The optimum condition for each factor affecting ferulic acid production is temperature at 26 °C, agitation speed at 150 rpm, pH value at pH 9, time at 1-day and inoculum percentage at 2%.

References

Abraham RE, Verma ML, Barrow CJ, Puri M (2014) Suitability of magnetic nanoparticle immobilised cellulases in enhancing enzymatic saccharification of pretreated hemp biomass. Biotechnol Biofuels 7(1):90

Bisswanger H (2014) Enzyme assays. Perspect Sci 1(1–6):41–55

Esteban-Torres M, Reverón I, Mancheño JM, de Las Rivas B, Muñoz R (2013) Characterization of a feruloyl esterase from Lactobacillus plantarum. Appl Environ Microbiol 79(17):5130–5136

Faulds CB, DeVries RP, Kroon PA, Visser J, Williamson G, de Vries RP, … Williamson G (1997) Influence of ferulic acid on the production of feruloyl esterases by Aspergillus niger. FEMS Microbiol Lett 157(2): 239–244

Gopalan N, Rodríguez-Duran LV, Saucedo-Castaneda G, Nampoothiri KM (2015) Review on technological and scientific aspects of feruloyl esterases: a versatile enzyme for biorefining of biomass. Bioresour Technol 193:534–544

He F, Zhang S, Liu X (2015) Immobilization of feruloyl esterases on magnetic nanoparticles and its potential in production of ferulic acid. J Biosci Bioeng 120(3):330–334

Huang S-H, Liao M-H, Chen D-H (2003) Direct binding and characterization of lipase onto magnetic nanoparticles. Biotechnol Prog 19(3):1095–1100

Khushairi ZA, Yussof HW, Rodzri NA, Abu Samah R, Zainol N (2016) A factorial analysis study on factors contribution to ferulic acid production from oil palm frond waste. Jurnal Teknologi 78(10):147–152

Kühnel S, Pouvreau L, Appeldoorn MM, Hinz SWA, Schols HA, Gruppen H (2012) The ferulic acid esterases of Chrysosporium lucknowense C1: purification, characterization and their potential application in biorefinery. Enzym Microb Technol 50(1):77–85

Li J, Cai S, Luo Y, Dong X (2011) Three feruloyl esterases in Cellulosilyticum ruminicola H1 act synergistically to hydrolyze esterified polysaccharides. Appl Environ Microbiol 77(17):6141–6147

Martinek RG (1969) Practical clinical enzymology. J Am Med Technol 31(2):161–186

McMurrough I, Madigan D, Donnelly D, Hurley J, Doyle A, Hennigan G, ..., Smyth MR (1996) Control of ferulic acid and 4-vinyl guaiacol in brewing. J Ins Brew 102(5): 327–332

Mohanasrinivasan V, Diksha D, Suganthi V, Selvarajan E, Subathra Devi C (2014) Process optimization for enhanced production of??-amylase by submerged (SmF) and solid state fermentation (SSF) using Bacillus pumilus VITMDS2. Res J Pharm Biol Chem Sci 5(2):1784–1800

Ou S, Zhang J, Wang Y, Zhang N (2011) Production of feruloyl esterase from aspergillus niger by solid-state fermentation on different carbon sources. Enzym Res 2011:848939

Ralph J, Quideau S, Grabber JH, Hatfield RD (1994) Identification and synthesis of new ferulic acid dehydrodimers present in grass cell walls. J Chem Soc Perkin Trans 1(23):3485–3498

Rodríguez Porcel EM, Casas López JL, Sánchez Pérez JA, Fernández Sevilla JM, Chisti Y (2005) Effects of pellet morphology on broth rheology in fermentations of Aspergillus terreus. Biochem Eng J 26(2–3):139–144

Shahzadi T, Anwar Z, Iqbal Z, Anjum A, Aqil T (2014) Induced production of exoglucanase, and β-glucosidase from fungal co-culture of. Adv Biosci Biotechnol 5(April):426–433

Silva JPA, Mussatto SI, Roberto IC (2010) The influence of initial xylose concentration, agitation, and aeration on ethanol production by Pichia stipitis from rice straw hemicellulosic hydrolysate. Appl Biochem Biotechnol 162(5):1306–1315

Szwajgier D, Pielecki J, Targoński Z, Brew JI (2005) The release of ferulic acid and feruloylated oligosaccharides during wort and beer production. J Inst Brew 111(January):372–379

Tsang SC, Yu CH, Gao X, Tam K (2006) Silica-encapsulated nanomagnetic particle as a new recoverable biocatalyst carrier. J Phys Chem B 110(34):16914–16922

Wang X, Geng X, Egashira Y, Sanada H (2004) Purification and characterization of a feruloyl esterase from the intestinal Bacterium Lactobacillus acidophilus. Appl Environ Microbiol 70(4):2367–2372

Xun E, Lv X, Kang W, Wang J, Zhang H, Wang L, Wang Z (2012) Immobilization of pseudomonas fluorescens lipase onto magnetic nanoparticles for resolution of 2-octanol. Appl Biochem Biotechnol 168(3):697–707

Chapter 6
Killer Yeast, a Novel Biological Control of Soilborne Diseases for Good Agriculture Practice

Azzam Aladdin, Julián Rafael Dib, Roslinda Abd. Malek, and Hesham A. El Enshasy

Abstract *Aspergillus niger* (*A. niger*) causes a disease called black mold on certain fruits and vegetables such as grapes, apricots, onions, and peanuts and is a common contaminant of food. Containment of this disease can reduce the amount of fruits, vegetables, and foods to be discarded, hence reducing the amounts of agricultural waste generated. Chemical control of *A. niger* has been partially successful, and fungicides are commonly used in the management of black mold. However, the risk of the establishment of resistant *Aspergillus* strains is considerable. Biocontrol, a nonhazardous alternative to the use of chemical fungicides, involves the use of biological processes to reduce crop loss and various microorganisms. Since it was first reported, the killer phenomenon in yeasts has been extensively studied in several genera and species, and its importance is gaining further recognition by industrialists. The food and beverage industries were among the first to explore the ability of toxin-producing yeasts to kill other fungus.

6.1 Introduction

The development of environmentally friendly alternatives to counter the extensive use of chemical pesticides is one of the biggest ecological challenges facing microbiologists and plant pathologists for combatting crop diseases. Biological control

A. Aladdin • R.A. Malek • H.A. El Enshasy (✉)
Institute of Bioproduct Development (IBD), Universiti Teknologi Malaysia,
Johor Bahru, Johor, Malaysia
e-mail: henshasy@ibd.utm.my

J.R. Dib
Planta Piloto de Procesos Industriales Microbiológicos (PROIMI-CONICET),
Tucumán, Argentina

Instituto de Microbiología, Facultad de Bioquimica, Quimica y Farmacia, Universidad
Nacional de Tucumán, Tucumán, Argentina

Z.A. Zakaria (ed.), *Sustainable Technologies for the Management of Agricultural Wastes*, Applied Environmental Science and Engineering for a Sustainable Future, https://doi.org/10.1007/978-981-10-5062-6_6

agents (BCAs) involve harnessing disease-suppressive microorganisms to improve plant health (Simsek 2011; Hornby 1990). Among plant pathogens, soilborne pathogens are considered to be more common than seed-borne or airborne diseases in the production of many crops and accounted for 10–20% of yield losses every year (Greenfield and Southgate 2003). Soilborne diseases suppression by BCA is the sustained manifestation of interactions among the plant, the pathogen, the BCA, the microbial community on and around the plant, and the physical environment (Hornby 1990). BCAs of soilborne diseases are often perceived to have generally beneficial impacts on the environment compared to conventional control of soilborne plant pathogens (Gomiero et al. 2008; Aldanondo and Almansa 2009). Good organic agriculture practices are regulated internationally by Codex Alimentarius Guidelines by the World Health Organization (WHO), United Nations Food and Agricultural Organization (FAO), and the International Federation of Organic Agriculture Movements (IFOAM). Briefly, four main principles are based on Willer et al. (2010) as follows:

1. *In health*: intended to produce high quality food without using mineral fertilizers and synthetic pesticides.
2. *In ecology*: should fit the cycles and balances in nature without exploiting it by using local resources, recycling, reuse, and efficient management of materials and energy.
3. *In fairness*: should provide good quality of life, contribute to food sovereignty, reduce poverty, enhance animal well-being, and take future generations into account.
4. *In care*: high responsibility and precaution have to be applied before adopting novel biological control technologies for organic agriculture, and significant risks should be prevented by rejecting unforeseeable technologies.

BCA of soilborne diseases is especially complex because these diseases come about in the dynamic environment at the interface of root and soil known as the rhizosphere, which is determined as the region surrounding a stem that is impressed by it (Chaube and Pundhir 2005). The rhizosphere is typified by rapid change, intense microbial activity, and high populations of bacteria compared with non-rhizosphere soil (Mehta et al. 2012). Plants release metabolically active cells from their sources and deposit as much as 20% of the carbon allocated to sources in the rhizosphere, suggesting a highly developed relationship between the plant and rhizosphere microorganisms (Paterson et al. 1997). The rhizosphere is subject to dramatic changes in a short temporal scale-rain events and daytime drought can result in variations in water potential, salt concentration, osmotic potential, pH, and soil particle structure (Abd-Elgawad et al. 2010). Over longer temporal scales, the rhizosphere can change due to root growth, interactions with other soil biota, and weathering processes (Young et al. 1998). It is the active nature of the rhizosphere that makes it an interesting setting for the interactions that lead to disease and a BCA of disease (Lugtenberg and Leveau 2007). There are various reports on the potential use of rhizosphere-associated particular bacteria species or mycorrhizal fungi in stimulating plant growth and BCA for multiple plant pathogens caused by soilborne (Pal and Gardener 2006; Banerjee et al. 2006; Johansson et al. 2004). The role of other microbial species, including yeasts, has received less attention (Nassar et al. 2005).

Yeasts are considered as fungi with vegetative states that predominantly reproduced by budding or fission, in vegetative phase growing mainly as single cells (Kreger 2013). They include ascomycetous and basidiomycetous yeasts. Killer yeast (KY) is a strain of yeast cells that contains toxic proteins which are lethal to receptive cells (Oro et al. 2016). To date, it has been known that under competitive conditions, the killer phenomenon offers a considerable advantage to these yeast strains against other sensitive microbial cells in their ecological niches (Liu et al. 2015). As a consequence of their nutritional preference, yeast populations are generally an order of magnitude higher in the rhizosphere as opposed to the bulk soil (Mašínová et al. 2017). A diverse range of yeasts exhibit plant growth-promoting characteristics, including pathogen inhibition (Fu et al. 2016), phytohormone production (Nassar et al. 2005), phosphate solubilization (Alonso et al. 2008), nitrogen and sulfur oxidation (Falih and Wainwright 1995), siderophore production (Sansone et al. 2005), and stimulation of mycorrhizal root colonization (Alonso et al. 2008). Recently, several researchers have reported that KYs and their toxins may be used as novel biological agents to control soilborne diseases (Amprayn et al. 2012; Botha 2011; Yuliar et al. 2015).

6.2 Soilborne Plant Pathogens

Soilborne pathogens belong to different classes of prokaryotic and eukaryotic microorganisms. They exist in the soil for short or long periods and survive on plant residues or as resting organisms until obtaining their growth requirements through stimuli such as root exudates. At that time they are able to compete with other microorganisms and start penetrating the root system (de Boer et al. 2005). They either stay inside the plants and consequently cause complete plant death or act out the plants to infect other parts of the root or other instills roots. Plants infected by soilborne diseases suffer from different known plant diseases such as root rot, root blackening, wilt, stunting, or seedling damping-off (Noble and Coventry 2010). The most prevalent soilborne pathogen *Fusarium oxysporum* causes serious losses in protected agricultural production areas all over the world (Baysal et al. 2013). Table 6.1 listed some of the most prevalent soilborne pathogenic species that infected different plants.

Table 6.1 Soilborne pathogenic species of different plants

Pathogens	Plant	Disease	References
Aphanomyces euteiches	Snap bean	Root rot	Stone et al. (2003)
Colletotrichum coccodes	Tomato	Anthracnose	Abbasi et al. (2002)
Erwinia tracheiphila	Cucumber	Bacterial wilt	Huelsman and Edwards (1998)
Fusarium oxysporum	Carnation	Wilt	Pera and Filippi (1987)
Phytophthora capsici	Pepper	Root rot	Kim et al. (1997)
Phytophthora fragariae	Raspberry	Crown rot	Noble and Coventry (2010)
Phytophthora nicotianae	Citrus	Root rot	Widmer et al. (1999)
Pythium ultimum	Cucumber	Damping-off	Fuchs (2002)
Rhizoctonia solani	Pea, lettuce	Damping-off	Fuchs (2002), Lewis et al. (1992)
Sclerotinia sclerotiorum	Onion	White rot	Coventry et al. (2002)
Verticillium dahliae	Potato	Early dying	La Mondia et al. (1999)
Xanthomonas campestris	Tomato	Bacterial spot	Abbasi et al. (2002)

6.3 Suppression with Compost

Compost prepared from a range of organic matters is already used worldwide for suppressing pathogens on organic and conventional field crops (Martin and Ramsubhag 2015). The advantages of this approach are based on the role of composts for the improvement of physical, chemical, and biological properties of soil, which can have positive effects on soilborne pathogens. The degradation of composts in soil can directly bear on the viability and endurance of a pathogen species by restricting available nutrients and releasing natural chemical substances with varying inhibitory properties (Table 6.2). On the other hand, carbon released during the degradation of composts leads to increasing soil microbial activity and thereby raises the likelihood of competition effects in the field (Bailey and Lazarovits 2003). Organic amendments to the soil have been evidenced to stimulate the natural processes of microorganisms that are antagonistic to diseases (Akhtar and Malik 2000). In addition, organic amendments often contain biologically active molecules such as vitamins, growth regulators, and toxins, which can affect soil microorganisms. For example, plant resistance against the bacterial wilt pathogen was enhanced through the augmented activities of ascorbate peroxidase, monodehydroascorbate reductase, dehydroascorbate reductase, and glutathione reductase following the application of compost (Youssef and Tartoura 2013).

6.4 Biological Control

Growing awareness about the negative impacts of extensive use of chemicals on environment, microbial biodiversity, animal health, and human health has sustain the interest in finding an alternative solution to the use of chemicals in biological

Table 6.2 Composted materials to suppress soilborne plant pathogens

Compost	Pathogen	Disease	Crop	References
Cotton gin trash	*Pythium arrhenomanes*	Root rot	Sugar cane	Dissanayake and Hoy (1999)
Eucalyptus bark	*Phytophthora nicotianae*	Root rot	Waratah	Hardy and Sivasithamparam (1991)
Green waste	*Fusarium culmorum*	Foot rot	Winter wheat	Tilston et al. (2002)
Green waste	*Gaeumannomyces graminis*	Take all	Winter wheat	Tilston et al. (2002)
Green waste	*Phoma medicaginis*	Black stem	Garden pea	Tilston et al. (2002)
Green waste	*Plasmodiophora brassicae*	Clubroot	Chinese cabbage	Tilston et al. (2002)
Green waste	*Pseudocercosporella herpotrichoides*	Eyespot	Winter wheat	Tilston et al. (2002)
Kitchen, green wastes	*Mycosphaerella pinodes*	Foot rot	Pea	Schuler et al. (1993)
Leaf waste, municipal waste	*Pythium myriotylum*	Damping-off	Cucumber	Ben-Yephet and Nelson (1999)
Municipal waste	*Fusarium oxysporum* f. sp. *lini*	Wilt	Flax	Serra-Wittling et al. (1996)
Municipal waste	*Phytophthora nicotianae*	Root rot	Citrus	Widmer et al. (1998)
Municipal waste	*Pythium aphanidermatum*	Damping-off	Cucumber	Ben-Yephet and Nelson (1999)
Poultry manure	*Phytophthora cinnamomi*	Root rot	Lupin	Aryantha et al. (2000)
Sewage sludge	*Aphanomyces euteiches*	Root rot	Pea	Lumsden et al. (1983)
Sewage sludge	*Fusarium oxysporum* f. sp. *melonis*	Wilt	Melon	Lumsden et al. (1983)
Sewage sludge	*Fusarium solani*	Foot rot	Pea	Lumsden et al. (1983)
Sewage sludge	*Phytophthora capsici*	Crown rot	Pepper	Lumsden et al. (1983)
Sewage sludge	*Pythium aphanidermatum*	Damping-off	Bean	Lumsden et al. (1983)
Sewage sludge	*Pythium myriotylum*	Blight	Bean	Lumsden et al. (1983)
Sewage sludge, bark	*Fusarium oxysporum* f. sp. *basilici*	Wilt	Basil	Ferrara et al. (1996)
Sewage sludge, bark	*Rhizoctonia solani*	Damping-off	Basil, bean	Ferrara et al. (1996)

(continued)

Table 6.2 (continued)

Compost	Pathogen	Disease	Crop	References
Tomato crop west	*Fusarium oxysporum* f. sp. *radicis-lycopersici*	Crown rot	Tomato	Cheuk et al. (2003)
Veg., fruit, green waste	*Pythium macrosporum*	Root	Bulbous iris	Van Os and van Ginkel (2001)
Vegetable waste	*Sclerotium cepivorum*	White rot	Onion	Coventry et al. (2001)
Yard waste	*Pythium ultimum*	Damping-off	Cucumber	Noble and Coventry (2010)

control (Whipps 2001). Some of the advantages of using biological control agents (BCAs) are as follows (Whipps and Gerhardson 2007; Quimby et al. 2002):

1. Reduced input of nonrenewable resources
2. Potentially self-sustaining
3. Long-term disease suppression in an environmentally friendly manner

The mechanisms employed by BCAs (dominated by bacteria, 90%, and fungi, 10%) are sustained by various interactions such as competition for nutrients and space, antibiosis, parasitism, and induced systemic resistance (Lucas 2009). It has been first reported that most patented BCAs belong to specific classes of bacteria (Montesinos 2003). Only recently fungus especially yeast have been used as biofertilizer and BCAs (Zhang 2000; Wang et al. 2016).

6.4.1 Bacteria

Bacteria are successfully used as BCAs and can have many positive effects on the plants (Emmert and Handelsman 1999). Since 1990, the biological pesticides Cedomon® and Cerall® have replaced almost 1.5 million liters of synthetic chemical fungicides in cereal production (Hökeberg 2005). To date, virulent strains of *Bacillus* spp., *Streptomyces* spp., *Pseudomonas* spp. and *Ralstonia* spp., *Acinetobacter* spp., *Paenibacillus* spp., and *Burkholderia* spp. have been reported with BCA properties (Yuliar et al. 2015). *Bacillus subtilis* has been one of the most commonly used and well-studied organisms by many researchers. The rhizobacterium *B. subtilis* has an average of 4–5% of its genome devoted to antibiotic synthesis such as bioactive peptides with great potential to use BCAs (Stein 2005; Kim et al. 2004). *B. subtilis* has shown particular utility in soil recovery (Amani et al. 2010), remediation of soil contaminated by heavy metals (Pacwa et al. 2011), and biocontrol against phytopathogens (Igo 1983). However, most of the known antifungal agents produced by *B. subtilis* are polypeptides (Stein 2005), including iturins A-E and bacillomycins D, F, and L (Bent 1999). These antifungal peptides inhibit the growth of a large number of fungi including *Aspergillus* spp., *Penicillium* spp., and *Fusarium* spp. especially *F. oxysporum* (Zhao and Teixeira 2006;

Munimbazi and Bullerman 1998; Baysal et al. 2013). Awad et al. (2014) reported that *Streptomyces glauciniger* is an active chitinase producer strain and exhibited BCA activity against *F. oxysporum*.

6.4.2 Fungi

The development of fungi for the biocontrol of soilborne pathogens has received a significant interest in recent years (Table 6.3). For example, *Trichoderma* spp. are very effective biological control agent for plant disease management especially for soilborne pathogens. It is a free-living fungus that is common in soil and root rhizosphere. It is highly interactive in root, soil, and foliar environments. It reduces growth, survival, or infections caused by pathogens by different mechanisms like competition, antibiosis, mycoparasitism, hyphal interactions, and enzyme secretion (Zaidi and Singh 2013; Thangavelu and Mustaffa 2012). The *Trichoderma* spp. may suppress the growth of the pathogen population in the rhizosphere through competition and thus reduce pathogen development. It produces antibiotics and toxins such as trichothecene and a sesquiterpene, trichodermin, which have a direct effect on other organisms (Daguerre et al. 2014). The antagonist *Trichoderma hyphae* either grow along the host hyphae or coil around it and secrete different lytic enzymes such as chitinase, glucanase, and pectinase that are involved in the process of mycoparasitism (Al-Naemi et al. 2016).

6.4.3 Killer Yeast

Killer yeast (KY) is considered to be one of the most promising BCA for more rational and safe crop management practices. It has been documented that the application of three rhizosphere yeasts, namely, *Rhodotorula glutinis* (*R. glutinis*), *Candida valida* (*C. valida*), and *Trichosporon asahii* (*T. asahii*) obtained from sugar beet rhizosphere, individually or in combination, significantly reduced postemergence damping-off of seedlings and crown and root rots of mature sugar beet caused by *Rhizoctonia solani* (*R. solani*) under special conditions (El-Tarabily 2004). This work showed different mechanisms of KY activity against pathogen of *R. solani*. *C. valida* produced β-1,3-glucanase or diffusible antifungal metabolites and degraded the hyphae of *R. solani* in vitro, causing hyphal plasmolysis and lysis of cell walls. *R. glutinis* produced only inhibitory volatiles, whereas *T. asahii* produced only diffusible antifungal metabolites, both inhibiting the vegetative growth of *R. solani* in vitro (El-Tarabily 2004). The specific mode of antagonistic activity of each of the three species indicates that individual species of the yeasts evaluated has the potential to employ different mechanisms to suppress a single pathogen, as with bacteria, actinomycetes, and filamentous fungi (Whipps 2001). The three yeast species did not inhibit each other, and in fact a combination was the most effective treatment for

Table 6.3 Fungi developed or being developed for the biological control of pathogens (Perez et al. 2016)

Fungus (BCAs)	Disease	Commercial name	Company
Candida oleophila strain O	Pome fruits diseases	Nexy	BioNext sprl, USA
Gliocladium catenulatum	Many plant diseases	Primastop	Kemira, Agro Oy, Finland
Phlebiopsis(Peniophora) gigantea	*Heterobasidium annosus*	Mycoparasites Rotstop	Kemira Agro Oy, Finland
Trichoderma harzianum, *T. polysporum*	Fungi causing wilt and wood decay	Binab T	Bio-Innovation, Sweden
T. harzianum	*Rhizoctonia solani*, *Sclerotium rolfsii*, *Pythium*	Trichoderma 2000	Mycontrol (EfA1)Ltd., Israel
T. harzianum	Wide range of fungal diseases	Trichopel	Agrimm Technologies Ltd., New Zealand
T. harzianum	*Pythium*, *Rhizoctonia*, *Fusarium*, *Sclerotina*	T-22 and T-22HB Bio-Trek, RootShield	BioWorks (TGT Inc) Geneva, USA
T. viride	*Chondrostereum purpureum* and soil and foliar pathogens	Trichodowels, Trichoject, Trichoseal, and others	Agrimm Technologies Ltd., New Zealand
T. harzianum	*Botrytis cinerea* and other fungal diseases	Trichodex	Makhteshim-Agan, Several European companies, e.g., DeCeuster, Belgium
Gliocladium virens	Damping-off and root pathogens	SoilGard	ThermoTrilogy, USA
Coniothyrium minitans	*Sclerotinia species*	Contans WG	Prophyta, Germany
Ampelomyces quisqualis	Powdery mildews	AQ10 biofungicide	Ecogen Inc., USA
Cryptococcus albidus	*Penicillium spp.* and *Botrytis spp.*	Yield plus	Anchor Yeast, S. Africa
Candida oleophila	*Penicillium spp.* and *Botrytis spp.*	Aspire	Ecogen Inc. USA
Fusarium oxysporum	*Fusarium oxysporum*	Fusaclean	Natural Plant Protection, France
F. oxysporum	*F. oxysporum* and *F. moniliforme*	Biofox C	SIAPA, Italy
Pythium oligandrum	*Pythium ultimum*	Polygandron Polyversum	Plant Protection Institute, Slovak Republic

protecting the sugar beet from pathogens in the glasshouse experiments (El-Tarabily 2004). From the findings obtained in works carried out in Universiti Teknologi Malaysia, two KY strains (*Pichia fermentans* and *Candida catenulate*) exhibited strong inhibitory effect against *F. oxysporum* in vitro (unpublished data).

The application of mixtures of antagonists enhanced biological control in comparison to individual candidates (Whipps 2001). It has also being reported that the application of *Candida glabrata*, *C. maltosa*, *C. slooffiae*, *Rhodotorula rubra*, and *Trichosporon cutaneum*, applied individually or as a mixture, significantly reduced the incidence of late-wilt disease of maize caused by *Cephalosporium maydis* (El-Mehalawy et al. 2004). The mode of action of these yeasts, tested in vitro, was found to be the production of antifungal diffusible metabolites and cell wall-degrading enzymes, including chitinase and β-1,3-glucanase. In addition, the rhizosphere yeasts *Saccharomyces unispora* and *Candida steatolytica* significantly reduced the incidence of wilt disease of beans caused by *F. oxysporum* through the production of antifungal diffusible metabolites.

6.4.4 Antagonism Mechanisms

Several mechanisms have been reported in the biocontrol activity of KYs against fungal soilborne pathogens which include competition for nutrients and space (Janisiewicz et al. 2000), production of cell wall lysis enzymes such as β-1,3-glucanase and chitinase (Urquhart and Punja 2002), mycoparasitism (Wisniewski et al. 1991), and production of antifungal diffusible and volatile metabolites (Masih et al. 2001). Therefore, understanding the antagonism mechanisms of the antagonists among KYs will help in the improvement of their performance resulting from the enhancement of their effectiveness as BCAs and in the development of criteria for rapid screening of superior BCAs.

6.5 Competition for Space and Nutrients

Several investigations have been carried out in relation to competition for space and nutrients. It has been suggested by Arras et al. (1998) that yeast colonies in close association with hyphae of fungal pathogens, indicating the attraction of these yeasts to the hyphal surfaces (hyphasphere) or to aggregates of mycelia where significant exudates or leakages from the filamentous pathogenic fungi could be expected to occur. El-Tarabily et al. reported that competition for nutrients could be particularly relevant in relation to sugary root exudates which may enhance the antagonistic activities of yeasts in relation to their competence to colonize plant roots (Fig. 6.1).

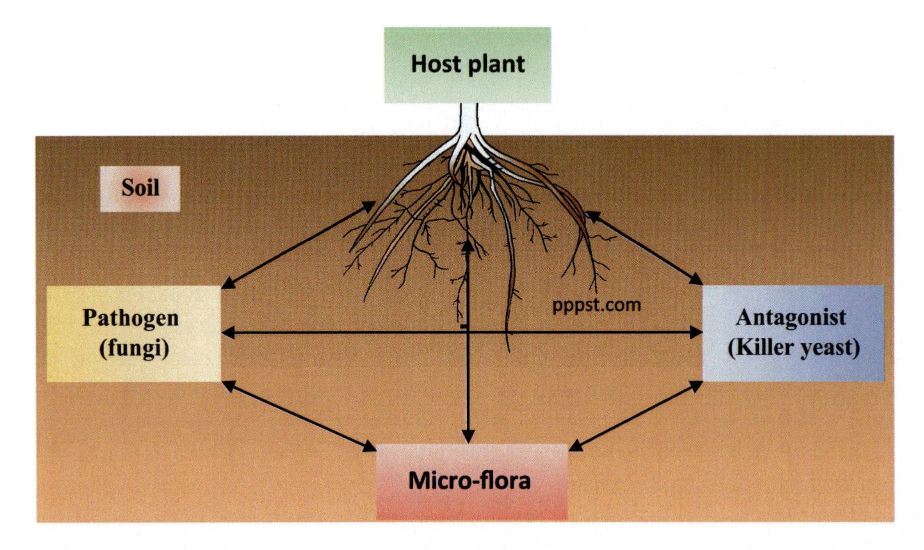

Fig. 6.1 Interactions between host roots, fungus pathogen, antagonist (killer yeast), and the microflora in the soil

6.5.1 Antibiosis

Various reports on KYs suggested the occurrence and the activity of antibiotics in the interaction with fungal pathogens. Urquhart and Punja (2002) reported that a fatty acid ester of the KY exhibited significant antifungal activity against soilborne fungal plant pathogens such as *F. oxysporum*, *Phoma* spp., and *Pythium aphanidermatum.* Suzzi et al. (1995) reported that KY belonging to *Saccharomyces* and *Zygosaccharomyces* genus inhibited the in vitro growth of soilborne pathogens such as *Rhizoctonia fragariae*, *Sclerotinia sclerotiorum*, and *Macrophomina phaseolina*. Figure 6.2 shows interactions between biocontrol plant growth-promoting rhizobacteria, plants, pathogens, and soil.

6.5.2 Cell Wall-Degrading Enzymes

It has been shown that KY have deleterious effect on exo- and endo-β-1,3-glucanases and chitinases (Castoria et al. 2001; Muccilli et al. 2013). Production of both endo- and exo-β-1,3-glucanases was stimulated by the presence of the cell wall preparation of *Botrytis cinerea* (Bauermeister et al. 2015). Figure 6.3 shows the model of the hydrolases action of β-1,3-glucanases and chitinases, the killer toxin against soilborne plant pathogen interactions. In fungi, wall extension is restricted to the hyphal tip and is thought to represent a delicate balance between synthesis and degradation of the main wall components, β-1,3-glucan (Mauch et al. 1988). However,

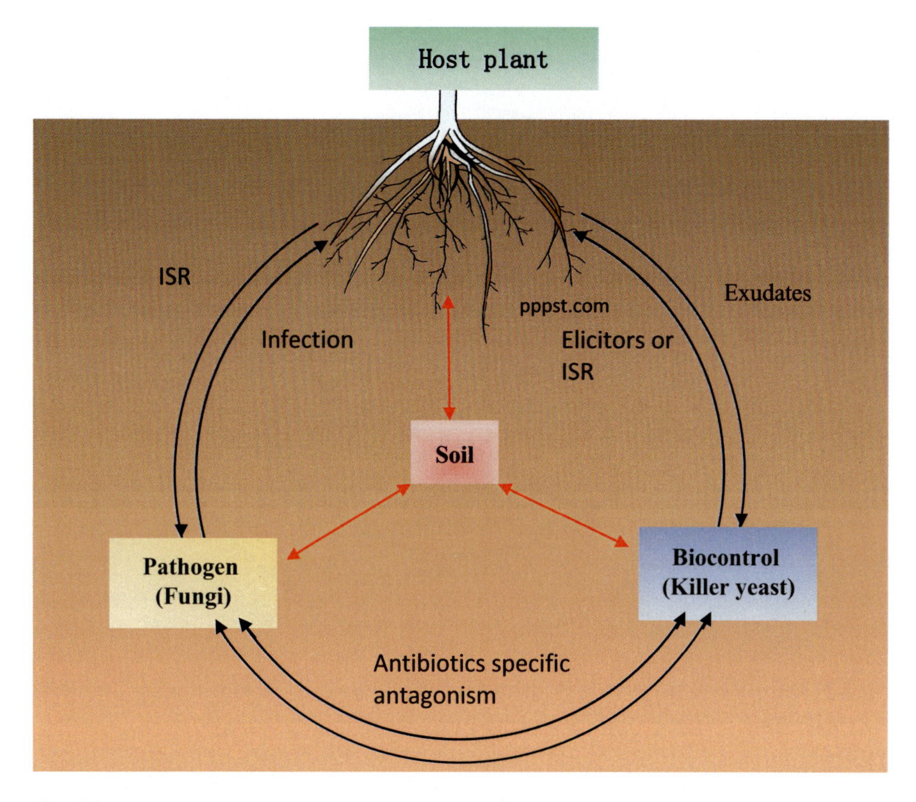

Fig. 6.2 Interaction between antibiotics from KY, plants, pathogens, and soil (Modified from Haas and Defago 2005)

exogenous application of chitinase and β-1,3-glucanase may disturb this balance. The requirement for β-1,3-glucanase to cause lysis indicates that most hyphal tips contain β-1,3-glucan. Hyphal walls of subapical regions and the walls of the fungal spores were resistant to the hydrolases, suggesting that β-1,3-glucan are protected by additional compounds at these locations (Ferrreira et al. 2012).

6.6 Conclusion

The central challenges of GAPs in conventional farming are to improve soil quality, reduce the usage of chemical pesticides and fertilizer, enhance plant growth, and protect biodiversity of soil microorganisms. This can be achieved by an integrated farming approach, in which disease controls are based mainly on preventive practices, such as versatile crop rotation and use of non-contaminated seeds, and pesticides are used only if particularly advantageous. The management of soil quality is a high priority for sustaining long-term productivity of the plant. Consequently,

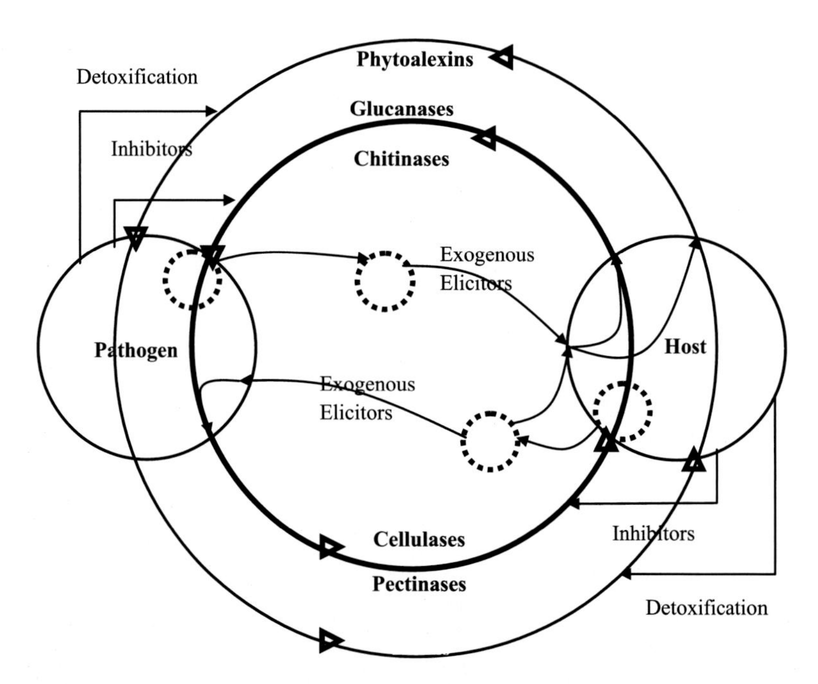

Fig. 6.3 The model of the hydrolases action of β-1,3-glucanases and chitinase as the killer toxin in plant pathogen interactions (Modified from Mauch et al. 1988)

practices that increase the soil organic matter content and the social organization of soil should be employed. Further study is needed to investigate the possibilities of mixing organic matter and BCAs as well as improving the land nutritional contents from the use of KY.

References

Abbasi P, Al-Dahmani J, Sahin F, Hoitink H, Miller S (2002) Effect of compost amendments on disease severity and yield of tomato in conventional and organic production systems. Plant Dis 86(2):156–161

Abd-Elgawad M, El-Mougy N, El-Gamal N, Abdel-Kader M, Mohamed M (2010) Protective treatments against soilborne pathogens in citrus orchards. J Plant Protect Res 50(4):477–484

Akhtar M, Malik A (2000) Roles of organic soil amendments and soil organisms in the biological control of plant-parasitic nematodes: a review. Bioresour Technol 74(1):35–47

Aldanondo AM, Almansa C (2009) The private provision of public environment: consumer preferences for organic production systems. Land Use Policy 26(3):669–682

Al-Naemi FA, Ahmed TA, Nishad R, Radwan O (2016) Antagonistic effects of Trichoderma harzianum isolates against Ceratocystis radicicola: pioneering a biocontrol strategy against black scorch disease in date palm trees. J Phytopathol 164(7–8):433–570

Alonso LM, Kleiner D, Ortega E (2008) Spores of the mycorrhizal fungus Glomus mosseae host yeasts that solubilize phosphate and accumulate polyphosphates. Mycorrhiza 18(4):197–204

Amani H, Mehrnia MR, Sarrafzadeh MH, Haghighi M, Soudi MR (2010) Scale up and application of biosurfactant from Bacillus subtilis in enhanced oil recovery. Appl Biochem Biotechnol 162(2):510–523

Amprayn KO, Rose MT, Kecskés M, Pereg L, Nguyen HT, Kennedy IR (2012) Plant growth promoting characteristics of soil yeast (*Candida tropicalis* HY) and its effectiveness for promoting rice growth. Appl Soil Ecol 61:295–299

Arras G, Cicco VD, Arru S, Lima G (1998) Biocontrol by yeasts of blue mould of citrus fruits and the mode of action of an isolate of *Pichia guilliermondii*. J Hortic Sci Biotechnol 73(3):413–418

Aryantha I, Cross R, Guest D (2000) Suppression of *Phytophthora cinnamomi* in potting mixes amended with uncomposted and composted animal manures. Phytopathology 90(7):775–782

Awad HM, El-Enshasy HA, Hanapi SZ, Hamed ER, Rosidi B (2014) A new chitinase-producer strain *Streptomyces glauciniger* WICC-A03: isolation and identification as a biocontrol agent for plants phytopathogenic fungi. Nat Prod Res 28(24):2273–2277

Bailey K, Lazarovits G (2003) Suppressing soil-borne diseases with residue management and organic amendments. Soil Tillage Res 72(2):169–180

Banerjee MR, Yesmin L, Vessey JK (2006) Plant-growth-promoting rhizobacteria as biofertilizers and biopesticides, Handbook of microbial biofertilizers. Food Products Press, New York, pp 137–181

Bauermeister A, Amador IR, Pretti CP, Giese EC, Oliveira AL, Alves da Cunha MA et al (2015) β-(1→3)-Glucanolytic yeasts from Brazilian grape microbiota: production and characterization of β-Glucanolytic enzymes by Aureobasidium pullulans 1WA1 cultivated on fungal mycelium. J Agric Food Chem 63(1):269–278

Baysal O, Lai D, Xu HH, Siragusa M, Caliskan M, Carimi F et al (2013) A proteomic approach provides new insights into the control of soil-borne plant pathogens by *Bacillus* species. PLoS One 8(1):e53182. https://doi.org/10.1371/journal.pone.0053182

Bent AF (1999) Applications of molecular biology to plant disease and insect resistance. Adv Agron 66:251–298

Ben-Yephet Y, Nelson EB (1999) Differential suppression of damping-off caused by *Pythium aphanidermatum*, *P. irregulare*, and *P. myriotylum* in composts at different temperatures. Plant Dis 83(4):356–360

de Boer W, Folman LB, Summerbell RC, Boddy L (2005) Living in a fungal world: impact of fungi on soil bacterial niche development. FEMS Microbiol Rev 29(4):795–811

Botha A (2011) The importance and ecology of yeasts in soil. Soil Biol Biochem 43(1):1–8

Castoria R, De Curtis F, Lima G, Caputo L, Pacifico S, De Cicco V (2001) *Aureobasidium pullulans* (LS-30) an antagonist of postharvest pathogens of fruits: study on its modes of action. Postharvest Biol Technol 22(1):7–17

Chaube H, Pundhir V (2005) Crop diseases and their management. PHI Learning Pvt. Ltd, New Delhi

Cheuk W, Lo KV, Branion R, Fraser B, Copeman R, Jolliffe P (2003) Applying compost to suppress tomato disease. Biocycle 44(1):50–51

Coventry E, Noble R, Whipps J (2001) Composting of onion and other vegetable wastes, with particular reference to Allium white rot. Rep Community Supported Agric 4862:1–95

Coventry E, Noble R, Mead A, Whipps J (2002) Control of Allium white rot (*Sclerotium cepivorum*) with composted onion waste. Soil Biol Biochem 34(7):1037–1045

Daguerre Y, Siegel K, Edel-Hermann V, Steinberg C (2014) Fungal proteins and genes associated with biocontrol mechanisms of soil-borne pathogens: a review. Fungal Biol Rev 28(4):97–125

Dissanayake N, Hoy J (1999) Organic material soil amendment effects on root rot and sugarcane growth and characterization of the materials. Plant Dis 83(11):1039–1046

El-Mehalawy AA, Hassanein NM, Khater HM, El-Din EK, Youssef YA (2004) Influence of maize root colonization by the rhizosphere actinomycetes and yeast fungi on plant growth and on the biological control of late wilt disease. Int J Agric Biol 6(4):599–605

El-Tarabily K (2004) Suppression of *Rhizoctonia solani* diseases of sugar beet by antagonistic and plant growth-promoting yeasts. J Appl Microbiol 96(1):69–75

Emmert EA, Handelsman J (1999) Biocontrol of plant disease: a gram positive perspective. FEMS Microbiol Lett 171(1):1–9
Falih A, Wainwright M (1995) Nitrification, S-oxidation and P-solubilization by the soil yeast Williopsis californica and by *Saccharomyces cerevisiae*. Mycol Res 99(2):200–204
Ferrara A;Avataneo M, Nappi P (1996) First experiments of compost suppressiveness to some phytopathogens. In The science of composting. Springer, Heidelberg, pp 1157–1160
Ferrreira RMSB, Freitas RFL, Monteiro SAVS (2012) Targeting carbohydrates: a novel paradigm for fungal control. Eur J Plant Pathol 133(1):117–140
Fu SF, Sun PF, Lu HY, Wei JY, Xiao HS, Fang WT et al (2016) Plant growth-promoting traits of yeasts isolated from the phyllosphere and rhizosphere of Drosera spatulata Lab. Fungal Biol 120(3):433–448
Fuchs J (2002) Practical use of quality compost for plant health and vitality improvement. In Microbiology of composting. Springer, Heidelberg, pp 435–444
Gomiero T, Paoletti M, Pimentel D (2008) Energy and environmental issues in organic and conventional agriculture. Crit Rev Plant Sci 27(4):239–254
Greenfield H, Southgate DA (2003) Food composition data: production, management, and use. Food & Agriculture Organization of the United Nations, Roma
Haas D, Defago G (2005) Biological control of soil-borne pathogens by fluorescent pseudomonads. Nat Rev Microbiol 3(4):307–319
Hardy GSJ, Sivasithamparam K (1991) Suppression of Phytophthora root rot by a composted Eucalyptus bark mix. Aust J Bot 39(2):153–159
Hökeberg M (2005) Development and registration of biocontrol products-experiences and perspectives gained from the bacterial seed treatment products Cedomon® and Cerall®, DIAS report, 77. Research Centre Flakkebjerg, Denmark
Hornby D (1990) Biological control of soil-borne plant pathogens: the centre for agriculture and bioscience international. CAB International, London
Huelsman M, Edwards C (1998) Management of disease in cucumbers (*Cucumis sativus*) and peppers (*Capsicum annum*) by using composts as fertility inputs. Food & Agriculture Organization of the United Nations, London
Igo N (1983) Survey of greenhouse management practices in Essex County, Ontario, in relation to Fusarium foot and root rot of tomato. Plant Dis 67(1):38–40
Janisiewicz W, Tworkoski T, Sharer C (2000) Characterizing the mechanism of biological control of postharvest diseases on fruits with a simple method to study competition for nutrients. Phytopathology 90(11):1196–1200
Johansson JF, Paul LR, Finlay RD (2004) Microbial interactions in the mycorrhizosphere and their significance for sustainable agriculture. FEMS Microbiol Ecol 48(1):1–13
Kim KD, Nemec S, Musson G (1997) Effects of composts and soil amendments on soil microflora and Phytophthora root and crown rot of bell pepper. Crop Prot 16(2):165–172
Kim P, Bai H, Bai D, Chae H, Chung S, Kim Y et al (2004) Purification and characterization of a lipopeptide produced by *Bacillus thuringiensis* CMB26. J Appl Microbiol 97(5):942–949
Kreger-van RNJW (2013) The yeasts: a taxonomic study. Elsevier, Tokyo, pp 45–65
LaMondia J, Gent M, Ferrandino F, Elmer W, Stoner K (1999) Effect of compost amendment or straw mulch on potato early dying disease. Plant Dis 83(4):361–366
Lewis J, Lumsden R, Millner P, Keinath A (1992) Suppression of damping-off of peas and cotton in the field with composted sewage sludge. Crop Prot 11(3):260–266
Liu GL, Chi Z, Wang GY, Wang ZP, Li Y, Chi ZM (2015) Yeast killer toxins, molecular mechanisms of their action and their applications. Crit Rev Biotechnol 35(2):222–234
Lucas J (2009) Plant pathology and plant pathogens. Wiley, Oxford, pp 233–249
Lugtenberg B, Leveau J (2007) 10 biocontrol of plant pathogens: principles, promises, and pitfalls. The rhizosphere: biochemistry and organic substances at the soil-plant interface. CRC Press, Boca Raton, pp 267–268
Lumsden R, Lewis J, Millner P (1983) Effect of composted sewage sludge on several soilborne pathogens and diseases. Phytopathology 73(11):1543–1548

Martin CCS, Ramsubhag A (2015) 18 potential of compost for suppressing plant diseases. Sustainable crop disease management using natural products. CAB International, Boston, pp 345–346
Masih E, Slezack-Deschaumes S, Marmaras I, Barka EA, Vernet G, Charpentier C et al (2001) Characterisation of the yeast *Pichia membranifaciens* and its possible use in the biological control of *Botrytis cinerea*, causing the grey mould disease of grapevine. FEMS Microbiol Lett 202(2):227–232
Mašínová T, Bahnmann BD, Větrovský T, Tomšovský M, Merunková K, Baldrian P (2017) Drivers of yeast community composition in the litter and soil of a temperate forest. FEMS Microbiol Ecol 93(2):fiw223. https://doi.org/10.1093/femsec/fiw223
Mauch F, Mauch-Mani B, Boller T (1988) Antifungal hydrolases in pea tissue II. Inhibition of fungal growth by combinations of chitinase and β-1, 3-glucanase. Plant Physiol 88(3):936–942
Mehta C, Gupta V, Singh S, Srivastava R, Sen E, Romantschuk M et al (2012) Role of microbiologically rich compost in reducing biotic and abiotic stresses. In: Microorganisms in environmental management. Springer, Heidelberg, pp 113–134
Montesinos E (2003) Development, registration and commercialization of microbial pesticides for plant protection. Int Microbiol 6(4):245–252
Muccilli S, Wemhoff S, Restuccia C, Meinhardt F (2013) Exoglucanase-encoding genes from three Wickerhamomyces anomalus killer strains isolated from olive brine. Yeast 30(1):33–43
Munimbazi C, Bullerman L (1998) Isolation and partial characterization of antifungal metabolites of *Bacillus pumilus*. J Appl Microbiol 84(6):959–968
Nassar AH, El-Tarabily KA, Sivasithamparam K (2005) Promotion of plant growth by an auxin-producing isolate of the yeast *Williopsis saturnus* endophytic in maize (Zea mays L.) roots. Biol Fertil Soils 42(2):97–108
Noble R, Coventry E (2010) Suppression of soil-borne plant diseases with composts: a review. Biocontrol Sci Tech 15(1):3–20
Oro L, Ciani M, Bizzaro D, Comitini F (2016) Evaluation of damage induced by Kwkt and Pikt zymocins against *Brettanomyces/Dekkera* spoilage yeast, as compared to sulphur dioxide. J Appl Microbiol 121:207–214
Pacwa PM, Płaza GA, Piotrowska SZ, Cameotra SS (2011) Environmental applications of biosurfactants. Int J Mol Sci 12(1):633–654
Pal KK, Gardener BM (2006) Biological control of plant pathogens. The Plant Health Instructor 2:1117–1142
Paterson E, Hall J, Rattray E, Griffiths B, Ritz K, Killham K (1997) Effect of elevated CO2 on rhizosphere carbon flow and soil microbial processes. Glob Chang Biol 3(4):363–377
Pera A, Filippi C (1987) Controlling of Fusarium wilt in carnation with bark compost. Biological Wastes 22(3):219–228
Perez MF, Contreras L, Garnica NM, Fernández-Zenoff MV, Farías ME, Sepulveda M et al (2016) Native killer yeasts as biocontrol agents of postharvest fungal diseases in lemons. PLoS One 11(10):e0165590. https://doi.org/10.1371/journal.pone.0165590
Quimby P, King L, Grey W (2002) Biological control as a means of enhancing the sustainability of crop/land management systems. Agric Ecosyst Environ 88(2):147–152
Sansone G, Rezza I, Calvente V, Benuzzi D, de Tosetti MIS (2005) Control of *Botrytis cinerea* strains resistant to iprodione in apple with rhodotorulic acid and yeasts. Postharvest Biol Technol 35(3):245–251
Schuler C, Pikny J, Nasir M, Vogtmann H (1993) Effects of composted organic kitchen and garden waste on *Mycosphaerella pinodes* (Berk. et Blox) Vestergr., causal organism of foot rot on peas (*Pisum sativum* L.). Biological Agriculture and Horticulture, London
Serra WC, Houot S, Alabouvette C (1996) Increased soil suppressiveness to *Fusarium* wilt of flax after addition of municipal solid waste compost. Soil Biol Biochem 28(9):1207–1214
Simsek EY (2011) The use of vermicompost products to control plant diseases and pests. In: Biology of earthworms. Springer, Berlin, pp 191–213

Stein T (2005) *Bacillus subtilis* antibiotics: structures, syntheses and specific functions. Mol Microbiol 56(4):845–857

Stone A, Vallad G, Cooperband L, Rotenberg D, Darby H, James R et al (2003) Effect of organic amendments on soilborne and foliar diseases in field-grown snap bean and cucumber. Plant Dis 87(9):1037–1042

Suzzi G, Romano P, Ponti I, Montuschi C (1995) Natural wine yeasts as biocontrol agents. J Appl Bacteriol 78(3):304–308

Thangavelu R, Mustaffa M (2012) Current advances in the Fusarium wilt disease management in banana with emphasis on biological control. INTECH, Shanghai, pp 274–287

Tilston E, Pitt D, Groenhof A (2002) Composted recycled organic matter suppresses soil-borne diseases of field crops. New Phytol 154(3):731–740

Urquhart E, Punja Z (2002) Hydrolytic enzymes and antifungal compounds produced by *Tilletiopsis* species, phyllosphere yeasts that are antagonists of powdery mildew fungi. Can J Microbiol 48(3):219–229

Van Os G, Van Ginkel J (2001) Suppression of pythium root rot in bulbous iris in relation to biomass and activity of the soil microflora. Soil Biol Biochem 33(11):1447–1454

Wang S, Liang Y, Shen T, Yang H, Shen B (2016) Biological characteristics of *Streptomyces albospinus* CT205 and its biocontrol potential against cucumber *Fusarium* wilt. Biocontrol Sci Tech 26(7):951–963

Whipps JM (2001) Microbial interactions and biocontrol in the rhizosphere. J Exp Bot 52(1):487–511

Whipps J, Gerhardson B (2007) Biological pesticides for control of seed-and soil-borne plant pathogens. Modern soil microbiology. CRC Press, Boca Raton, pp 479–501

Widmer T, Graham J, Mitchell D (1998) Composted municipal waste reduces infection of citrus seedlings by Phytophthora nicotianae. Plant Dis 82(6):683–688

Widmer T, Graham J, Mitchell D (1999) Composted municipal solid wastes promote growth of young citrus trees infested with *Phytophthora nicotianae*. Compost Science & Utilization 7(2):6–16

Willer H, Yussefi M, Sorensen N (2010) The world of organic agriculture. Statistics and emerging trends 2008. IFOAM, Earthscan, London

Wisniewski M, Biles C, Droby S, McLaughlin R, Wilson C, Chalutz E (1991) Mode of action of the postharvest biocontrol yeast, Pichia guilliermondii. I. Characterization of attachment to *Botrytis cinerea*. Physiol Mol Plant Pathol 39(4):245–258

Young IM, Blanchart E, Chenu C, Dangerfield M, Fragoso C, Grimaldi M et al (1998) The interaction of soil biota and soil structure under global change. Glob Chang Biol 4(7):703–712

Youssef SA, Tartoura KA (2013) Compost enhances plant resistance against the bacterial wilt pathogen *Ralstonia solanacearum* via up-regulation of ascorbate-glutathione redox cycle. Eur J Plant Pathol 137(4):821–834

Yuliar, Nion YA, Toyota K (2015) Recent trends in control methods for bacterial wilt diseases caused by *Ralstonia solanacearum*. Microbes Environ 30(1):1–11

Zaidi NW, Singh US (2013) 14 Trichodermain plant health management. Trichoderma: biology and applications. CAB International, London, p 230

Zhang L (2000) Biological fertilizer based on yeasts. US Patent, US6416983, September 5, 2000

Zhao J, Teixeira da Silva J (2006) Elicitor signal transduction leading to biosynthesis of plant defensive secondary metabolites. Floriculture, ornamental and plant biotechnology. Global Science Books, Ltd, Tokyo, pp 344–357

Chapter 7
Killer Yeasts as Biocontrol Agents of Postharvest Fungal Diseases in Lemons

María Florencia Perez, Ana Sofía Isas, Azzam Aladdin, Hesham A. El Enshasy, and Julián Rafael Dib

Abstract One of the main problems affecting the citrus industry worldwide is caused by fungal diseases at postharvest stage. This leads to huge amounts of lemon fruit being unnecessarily discarded which contributed to the overall generation of agricultural wastes. Synthetic fungicides are nowadays the major agents used to control diseases of fungal origin. However, long-term and uncontrolled usage may lead to environmental problems such as growing restrictions that are mainly due to its toxicity. Among biocontrol agents, killer yeasts appear as efficient candidates especially for combating fungal postharvest decay in lemons.

7.1 Introduction

After harvesting, fresh fruits are susceptible to be attacked by saprophytic pathogens or parasites due to their high water and nutrient content, and they have lost most of the intrinsic resistance that protects them during their growth on the tree. Moreover, its organic acid content is sufficient to produce a pH lower than 4.6 which mainly favors infection of fungal origin (Viñas 1990). Economic losses caused by postharvest diseases (PHD) are currently one of the main problems of global fruit

M.F. Perez • A.S. Isas
Planta Piloto de Procesos Industriales Microbiológicos (PROIMI-CONICET), Tucumán, Argentina

A. Aladdin • H.A. El Enshasy
Institute of Bioproduct Development, Universiti Teknologi Malaysia, Johor Bahru, Johor, Malaysia

J.R. Dib (✉)
Planta Piloto de Procesos Industriales Microbiológicos (PROIMI-CONICET), Tucumán, Argentina

Instituto de Microbiología, Facultad de Bioquímica, Química y Farmacia, Universidad Nacional de Tucumán, Tucumán, Argentina
e-mail: jdib@proimi.org.ar

Z.A. Zakaria (ed.), *Sustainable Technologies for the Management of Agricultural Wastes*, Applied Environmental Science and Engineering for a Sustainable Future, https://doi.org/10.1007/978-981-10-5062-6_7

Fig. 7.1 Green mold and blue mold are the most common postharvest diseases of lemons. Symptoms of blue mold (left) and a dual infection with blue and green mold (right)

horticulture (Harvey 1978; Kelman 1989), being the reduction of consumable units and the most evident loss at this stage. There is a large variety of PHD that affects different types of fruits. Among citrus fruits, the main ones are the "green mold" and "blue mold," produced by *Penicillium digitatum* and *P. italicum*, respectively (Fig. 7.1). To control PHD in citrus, the use of synthetic fungicides is the most widely used method, mainly due to its relative low cost, ease of application, and effectiveness. Among them, thiabendazole (TBZ) and imazalil (IMZ) received the most widespread use. However, since fungal populations are indeed biologically dynamic, the occurrence of resistant strains to these fungicides frequently occurs, hence limiting their efficacy (Sánchez-Torres and Tuset 2011). In addition, the growing public health and environmental concerns have resulted in the de-registration of some of the most effective synthetic fungicides (Makovitzki et al. 2007). Yeasts as biocontrol agents of postharvest pathogens have been especially investigated because of their advantages and potential (Droby et al. 2009; Kefialewa and Ayalewb 2008; Chanchaichaovivat et al. 2007; Janisiewicz et al. 2008; Wang et al. 2008; Liu and Tsao 2009; Hashem and Alamri 2009; Zhang et al. 2010; Rosa et al. 2010).

7.2 Postharvest Diseases in Citrus

After harvesting, the most important fungi affecting citrus fruits are *P. digitatum* and *P. italicum*, the former being the most common pathogen and having the highest reproductive activity (Vázquez et al. 1995; Garrán 1996; Ragone 1999). These phytopathogenic fungi are active when the fruits are at an advanced ripening stage or completely mature. Usually contamination takes place both on the field, while the fruit is still on the tree, during harvesting, transport, packing, handling, and preservation or during the period of distribution and sale in the markets. The main source of contamination is the conidia. Although conidia are of small size, produced in large quantities and transported by air currents to healthy fruits, they do not germinate until the fruit skin is damaged by some biotic (insects) or abiotic factors such as hail, wind, and harvest wounds (Tuset 1987; Cocco 2005). Conidia germination requires free water and nutrients. In the absence of nutrients, volatile compounds produced by damaged fruits stimulate germination (Eckert et al. 1984). Therefore, these fungi are pathogens of wounds and cause a soft and wet rot that rapidly deteriorates the organoleptic characteristics of the fruit (Tuset 1987).

7.3 Control of Postharvest Diseases in Citrus Fruits Through the Use of Synthetic Fungicides

As it was mentioned before, the utilization of synthetic fungicides is the most extensively used method to control PHD. TBZ, a product derived from benzimidazole, acts by avoiding the polymerization of the tubulin and thus inhibiting mitosis. The advantages of this product include its great effectiveness for wound fungi, able to penetrate through the cuticle and the epidermis inside the cortex in such a way that is able to stop the inactive infections or the young mycelial development, its great anti-sporulation activity against *Penicillium* spp., and its low toxicity. However, since fungitoxic activity is not exerted directly on the conidium, but rather on the germ tube or apices of growing young hyphae, resistant strains occur relatively easily (Tuset 1987; Garrán 1996). IMZ is an inhibitor of ergosterol biosynthesis, thus affecting the structure of cell membranes. This product is widely used in citrus packaging plants worldwide for the control of green and blue molds (Tuset 1987; Muller 2005). TBZ and IMZ are systemic fungicides that act on specific targets, and any mutation in the corresponding genes could develop resistance. In addition, both are ineffective (or less effective) against *Geotrichum* sp. (Suprapta et al. 1997; Mercier and Smilanick 2005). New fungicides have recently been registered for controlling citrus green mold such as pyrimethanil and fludioxonil, which belong to a new generation group that has been classified as "low risk" in 1998 by the United States Environmental Protection Agency (EPA). Their efficacy is similar to the known fungicides in use, and they have not shown fungi resistance issues. The mode of action of pyrimethanil is the blockade of protein synthesis through the inhibition of methionine biosynthesis. Fludioxonil blocks protein kinase, inhibiting the growth and development of fungi (Muller 2005; Smilanick et al. 2008; Velázquez et al. 2010).

7.4 Control of Postharvest Diseases in Citrus Fruits by Alternative Methods to Synthetic Fungicides

As a result of the massive and repeated use of fungicides, and in many cases the lack of knowledge and awareness about their correct utilization, several problems have been generated such as the emergence of resistant strains (Droby et al. 2002; Mercier and Smilanick 2005; Boubaker et al. 2009). In addition, its use is becoming increasingly restricted due to high residual toxicity, carcinogenicity, and long periods of degradation (Tripathi and Dubey 2003; Palou et al. 2008). Furthermore, in citrus wounds the pH decreases, and this dramatically alters the effectiveness of chemical fungicides since their neutral forms penetrate the membrane of pathogens and they are more toxic than ionized forms (López-García et al. 2003). On the other hand, lemon peel has many traditional uses, particularly in medicine for its antioxidant properties (Bocco et al. 1998) as well as in typical food dishes. However, the accumulation of residues of chemical fungicides mainly occurs in the peel, and its removal is quite difficult by simple washing. Therefore, the growing need for methods of low environmental impact and minimum risk for human health demands the constant development of novel products or processes as an alternative to fungicides.

7.5 Physical and Chemical Methods

Physical methods, such as different thermal or radiation treatments, and alternative chemical methods by the use of natural or synthetic chemicals with low residual and toxicological effects have been tested (Palou 2007). For example, Utama et al. (2002) demonstrated the efficacy of natural compounds as acetaldehyde, benzaldehyde, cinnamaldehyde, ethanol, benzyl alcohol, nerolidol, and 2-nonanone as volatile fungitoxicants for the protection of citrus fruit against *P. digitatum*. Carbonate and sodium bicarbonate are common food additives also used in the control of *P. digitatum* (Torres Leal et al. 2008; Carbajo 2011), which has few regulatory barriers, and it is recognized as safe by the Food and Drug Administration of the United States (FDA). However, although sodium salts have good control of citrus pathogens, they do not reach the efficiency of treatments such as curing (Meier 2005); thus their single application is not enough to control postharvest rots.

7.6 Biological Methods

On the other hand, biological methods also show great potential as an alternative method for the control of PHD in citrus (Droby et al. 2009). In a strict sense, the biological control of plant diseases is restricted to the controlled use of microorganisms that antagonizes pathogenic microorganisms. It can be defined as the direct or indirect manipulation on living agents (antagonist), the ones that naturally have the ability to control pathogens (Palou 2007; Palou et al. 2008). Interestingly, most of the antagonistic microorganisms are isolated from the surface of the fruits. It is thought that various mechanisms of action are involved in processes of biological control (Wisnieswski et al. 1991; Sharma et al. 2009; Jamalizadeh et al. 2011). These mechanisms are generally based on the ability of biocontrol agents to (1) adhere to specific sites, including both hosts and pathogenic cells (Wisnieswski et al. 2007), (2) colonize wounds and compete for nutrients, (3) secrete specific enzymes (Grevesse et al. 2003), (4) induce resistance (Yao and Tian 2005), (5) regulate population density at specific sites (McGuire 2000), (6) secrete antimicrobial substances (water-soluble or volatile), and (7) form biofilm on the inner surface of wounds (Giobbe et al. 2007). However, there is a potential limitation in the use of biocontrol agents: their adaptability to the conditions prevailing in each fruit and in the environments where they are stored. In this sense, the selection and use of native antagonist microorganisms isolated from the same environment in which the fruits are stored became an effective strategy to prevent PHD caused by phytopathogenic fungi. Indeed, there is also the consumer's fear that biological agents or toxins are introduced into the diet. However, since most of these agents are originally isolated from fruits and vegetables, they are indigenous to basic agricultural products (Droby et al. 2009). In addition, even when they are introduced in large numbers on the surface of a fruit, they can only survive and grow in very restricted sites (wounds). Furthermore, from the commercial point of view, the application of these biological methods may not always provide suitable levels of postharvest control. For this reason, in recent years, numerous investigations have been carried out with different antagonistic microorganisms, alone or in combination with physical treatments, in order to find an appropriate control method to replace the chemical synthetic fungicides or to reduce their doses (Janisiewicz and Korsten 2002).

7.6.1 Yeasts as Biocontrol Agents

Some of the advantages of using yeasts as biocontrol agents for postharvest pathogens are: (1) they show rapid growth in fermentors on economic substrates and, therefore, it is easy to produce in large quantities (Druvefors 2004); (2) they do not produce allergenic spores or mycotoxins, in contrast to filamentous fungi (Fan and Tian 2000); and (3) they have simple nutritional requirements that allow them to colonize dry areas for long periods of time (El-Tarabily and Sivasithamparam 2006). Additionally, yeast produces extracellular polysaccharides for own survival and to restrict the growth of pathogens (Sharma et al. 2009). For these reasons, the treatment of fruits with antagonistic yeasts is one of the best alternatives for the biological control of PHD (Janisiewicz and Korsten 2002). Regarding the control of postharvest fungal diseases in citrus through the use of yeasts, numerous studies have also been reported. Wilson and Chalutz (1989) observed that *Candida guilliermondii* and *Debaryomyces hansenii* were active against *P. digitatum* on grapefruit. Katz et al. (1995) developed a commercial product based on *Candida oleophila* as selective biofungicide for the control of postharvest citrus pathogens, whereas McGuire and Hagenmaier (1995) used this yeast mixed with wax applied to fruit during packaging. As it was already mentioned, there are several mechanisms of action in biological control, including the ability of the biocontrol agent to secrete antimicrobial substances. Within this group, there are yeasts with a "killer" phenotype, which have been described for the first time in *Saccharomyces cerevisiae* (Makower and Bevan 1963). These yeasts have the ability to secrete low molecular mass proteins or glycoprotein toxins that are lethal to susceptible cells (Coelho et al. 2007). These toxins, as they are active at pH values between 3 and 5.5 (Golubev and Shabalin 1994; Marquina et al. 2002), have an advantage over traditional fungicides that diminish their efficacy due to the reduction of pH values in the injured fruits (López-García et al. 2003). On the other hand, killer yeasts and their toxins have found applications in medicine as antifungal agents for the treatment of fungal infections in humans and animals (Magliani et al. 2008) and in biotechnology to eliminate undesirable microorganisms in industrial fermentation or food preservation (Sulo and Michalcakova 1992; Sulo et al. 1992; Lowes et al. 2000). Recently, 437 epiphytic yeast strains were isolated in our lab from citrus plants in Tucuman city (Argentina), which is known as one of the major producers and manufacturers of lemons in the world. From these 437 strains, 8.5% of the yeasts showed a *killer* phenotype. In vitro antagonistic activity of killer isolates against phytopathogenic fungi *P. digitatum*, *P. italicum*, and *Phomopsis citri* was evaluated. According to molecular identification, based on the 26S rDNA D1/D2 domain sequences analysis, strains were identified belonging to the genera *Saccharomyces*, *Wickerhamomyces*, *Kazachstania*, *Pichia*, *Candida*, and *Clavispora* (Perez et al. 2016). Subsequently, killer strains that had caused the maximum in vitro inhibition of *P. digitatum* were selected, and their ability to inhibit the fungus in in vivo assays in lemons was evaluated regarding extent and type of control and comparing with the protective effects of a commercial product based on *Candida oleophila* (unpublished results). Two strains of *Pichia* and one strain of *Wickerhamomyces* caused a significant growth inhibition ($p < 0.05$) of *P. digitatum* (Fig. 7.2). In the three cases, although no curative effect was observed in infected fruits, in the control by wounds, protection efficiencies up to 93.6% were determined (Perez et al. 2016). Similarly, an antagonistic activity higher than the commercial product in strains of the genus *Clavispora* and *Candida* was demonstrated (Perez et al. 2017). In addition, a strain of *Kazachstania exigua* and another of *Saccharomyces cerevisiae* efficiently inhibited the development of *P. italicum* in lemons (Perez et al. 2016). Interestingly, this strain of *Kazachstania exigua* also demonstrated antimicrobial effects against *Klebsiella pneumoniae*, *Escherichia coli*, and *Pseudomonas aeruginosa* (Perez et al. 2016).

Fig. 7.2 Biocontrol of green mold of lemons with killer yeast *Pichia fermentans* 27 (Perez et al. 2016). Pretreated lemons with killer yeast 27 (first row) and control fruits only inoculated with P. digitatum (second row)

Some of these yeast genera have also been described by other authors as biocontrol agents of phytopathogens in citrus. Platania et al. (2012) reported the biocontrol of *P. digitatum* on Tarocco oranges using killer yeasts of the species *Wickerhamomyces anomalus*. In addition, the exo-β-1,3-glucanase Panomycocin produced by *Pichia anomala* strain NCYC 434 showed 100% protection in fresh lemons against *P. digitatum* and *P. italicum* even after 5–7 days of incubation (Izgu et al. 2011). Other examples of yeasts as biocontrol agents of postharvest fungal diseases in citrus and their modes of action are shown in Table 7.1.

7.7 Conclusion

It is unrealistic to assume that microbial antagonists have the same fungicidal activity as fungicides. Thus, at the present, the use of postharvest biocontrol agents is constrained by the lack of consistent efficacy and the high level of control required at postharvest stage. Therefore, only a few products with high biocontrol potential have been made available on a commercial scale. Based on our findings, killer yeasts proved to be attractive candidates for the development of an effective, safe, and economical biological control agent to combat postharvest fungal infections of citrus. This would also act as an excellent alternative to the use of synthetic fungicides which ultimately anticipated to reduce the generation of agricultural waste.

Table 7.1 Antagonist yeasts used for controlling citrus postharvest diseases and their mode of action*

Yeast	Pathogen fungus	Mode of action	References
Pichia guilliermondii	*P. italicum*	Competition for nutrient and space	Arras et al. (1998)
Pichia anomala	*P. digitatum*	Competition for nutrient and space	Taqarort et al. (2008)
Pichia pastoris	*Geotrichum citri-aurantii*	Antibiosis	Ren et al. (2011)
Pichia membranaefaciens	*Penicillium* spp.	Induction of resistance	Luo et al. (2012).1
Candida saitoana	*P. digitatum*	Competition for nutrient and space	El-Ghaouth et al. (2000)
Candida famata	*P. digitatum*	Induction of resistance	Arras (1996)
Candida guilliermondii	*P. digitatum*	Induction of resistance	Arras (1996)
Candida oleophila	*P. digitatum*	Induction of resistance	Droby et al. (2002)
Candida sake	*P. digitatum*	ND	Droby et al. (1999)
Rhodotorula glutinis	*P. digitatum*	Competition for nutrient and space	Zheng et al. (2005)
Rhodosporidium paludigenum	*G. citri-aurantii*	Competition for nutrient and space	Liu et al. (2010)
Debaryomyces hansenii	*P. digitatum*	Competition for nutrient and space	Taqarort et al. (2008)
Cryptococcus laurentii	*P. italicum* *G. citri-aurantii*	Competition for nutrient and space	Zhang et al. (2005) and Liu et al. (2010)
Kloeckera apiculata	*P. italicum*	Competition for nutrient and space	Long et al. (2005, 2006)
Rhodotorula minuta, Candida azyma, Aureobasidium pullulans	*G. citri-aurantii*	Killer activity and hydrolytic enzyme production	Ferraz et al. (2016)
D. hansenii	*P. italicum*	ND	Hernández-Montiel et al. (2010)

*Modified according Talibi et al. 2014
ND not determined

References

Arras G (1996) Mode of action of an isolate of *Candida famata* in biological control of *Penicillium digitatum* in orange fruits. Postharvest Biol Technol 8:191–198

Arras G, De Cicco V, Arru S, Lima G (1998) Biocontrol by yeasts of blue mould of citrus fruits and the mode of action of an isolate of *Pichia guilliermondii*. J Hortic Sci Biotechnol 73:413–418

Bocco A, Cuvelier MA, Richard H, Berset C (1998) Antioxidant activity and phenolic composition of citrus peel and seed extracts. J Agric Food Chem 46:2123–2129

Boubaker H, Saadi B, Boudyach EH, Ait Benaoumar A (2009) Sensitivity of *Penicillium digitatum* and *Penicillium italicum* to imazalil and thiabendazole in Morocco. Plant Pathol J 4:152–158

Carbajo MS (2011) Sistemas alternativos a los fungicidas químicos para el control de *Penicillium digitatum* Sacc. en limón. Trabajo de tesis para optar al grado de Magíster en Producción Vegetal, Universidad de Buenos Aires, Argentina, p 124

Chanchaichaovivat A, Ruenwongsa P, Panijpan B (2007) Screening and identification of yeast strains from fruits and vegetables: potential for biological control of postharvest chilli anthracnose (*Colletotrichum capsici*). Biol Control 42:326–335

Cocco M (2005) Determinación de resistencia a fungicidas tradicionales en cepas de *Penicillium digitatum* y *Penicillium italicum* en distintas quintas y empaques de la región. Actas del II Seminario Internacional de Postcosecha de Cítricos. Ediciones INTA, pp 104–107

Coelho AR, Celli MG, Ono EYS, Wosiacki G (2007) *Penicillium expansum* versus antagonist yeasts with perspectives of application in biocontrol and patulin degradation. Braz Arch Biol Technol 50:725–733

Droby S, Porat R, Cohen L, Weiss B, Shapiro B, Philosoph-Hadas S, Meir S (1999) Suppressing green mold decay in grapefruit with postharvest jasmonate application. J Am Soc Hortic Sci 124:184–188

Droby S, Vinokur V, Weiss B, Cohen L, Daus A, Goldschmidt E, Porat R (2002) Induction of resistance to *Penicillium digitatum* in grapefruit by the yeast biocontrol agent *Candida oleophila*. Phytopathology 92:393–399

Droby S, Wisniewski M, Macarisin D, Wilson C (2009) Twenty years of postharvest biocontrol research: is it time for a new paradigm? Postharvest Biol Technol 52:137–145

Druvefors UÄ (2004) Yeast biocontrol of grain spoilage moulds – mode of action of *Pichia anomala*. Doctor's dissertation ISSN 1401-6249, IBSN 91-576-6493-5. Retrieved January 11, 2011, from Epsilon dissertations and graduate. Theses archive web site: http://diss-epsilon.slu.se:8080/archive/00000552/01/U%C3%84Dfin0.pdf

Eckert JW, Ratnayake M, Gutter Y (1984) Volatiles from wounded citrus fruits stimulate germination of *Penicillium digitatum* conidia. Phytopathology 74:793

El-Ghaouth A, Smilanick JL, Wilson CL (2000) Enhancement of the performance of *Candida saitoana* by the addition of glycochitosan for control of postharvest decay of apple and citrus fruit. Postharvest Biol Technol 19:249–253

El-Tarabily KA, Sivasithamparam K (2006) Potential of yeasts as biocontrol agents of soil-borne fungal plant pathogens and as plant growth promoters. Mycoscience 47:25–35

Fan Q, Tian SP (2000) Postharvest biological control of *Rhizopus* rot of nectarine fruits by *Pichia membranefaciens*. Plant Dis 84:1212–1216

Ferraz LP, da Cunha T, da Silva AC, Kupper KC (2016) Biocontrol ability and putative mode of action of yeasts against *Geotrichum citri-aurantii* in citrus fruit. Microbiol Res 188:72–79

Garrán SM (1996) Enfermedades durante la postcosecha. Manual para productores de naranjas y mandarinas de la región del río Uruguay. INTA 12:173–240

Giobbe S, Marceddu S, Scherm B, Zara G, Mazzarello VL, Budroni M, Migheli Q (2007) The strange case of a biofilm-forming strain of *Pichia fermentans*, which controls *Monilinia* brown rot on apple but is pathogenic on peach fruit. FEMS Yeast Res 7(8):1389–1398

Golubev W, Shabalin Y (1994) Mycocin production by the yeast *Cryptococcus humicola*. FEMS Microbiol Lett 119:105–110

Grevesse C, Lepoivre F, Jijakli MH (2003) Characterization of the exoglucanase-encoding gene aEXG2 and study of its role in the biocontrol activity of *Pichia anomala* strain K. Phytopathology 93:1145–1152

Harvey JM (1978) Reduction of losses in fresh market fruits and vegetables. Annu Rev Phytopathol 16:321–341

Hashem M, Alamri S (2009) The biocontrol of postharvest disease (*Botryodiplodia theobromae*) of guava (*Psidium guajava L.)* by the application of yeast strains. Postharvest Biol Technol 53:123–130

Hernández-Montiel LG, Ochoa JL, Troyo-Diéguez E, Larralde-Corona CP (2010) Biocontrol of postharvest blue mold (*Penicillium italicum Wehmer*) on Mexican lime by marine and citrus *Debaryomyces hansenii* isolates. Postharvest Biol Technol 56(2):181–187

Izgu DA, Kepekci RA, Izgu F (2011) Inhibition of *Penicillium digitatum* and *Penicillium italicum* in vitro and in planta with Panomycocin, a novel exo-β-1,3-glucanase isolated from *Pichia anomala* NCYC 434. Antonie Van Leeuwenhoek 99:85–91

Jamalizadeh M, Etebarian HR, Aminian H, Alizadeh A (2011) A review of mechanisms of action of biological control organisms against postharvest fruit spoilage. Bull OEPP/EPPO 41:65–71

Janisiewicz WJ, Korsten L (2002) Biological control of postharvest diseases of fruits. Annu Rev Phytopathol 40:411–441

Janisiewicz WJ, Saftner RA, Conway WS, Yoder KS (2008) Control of blue mold decay of apple during commercial controlled atmosphere storage with yeast antagonists and sodium bicarbonate. Postharvest Biol Technol 49:374–378

Katz H, Berkovitz A, Chalutz E, Droby S, Hofstein R, Karen-Tzoor M (1995) Compatibility of ecogens biofungicide aspire, a yeast based preparation, with other fungicides commonly used for the control of postharvest decay of citrus. Phytopathology 85:1123

Kefialewa Y, Ayalewb A (2008) Postharvest biological control of anthracnose (*Colletotrichum gloeosporioides*) on mango (*Mangifera indica*). Postharvest Biol Technol 50:8–10

Kelman A (1989) Introduction: the importance of research on the control of postharvest diseases of perishable food crops. Phytopathology 79:1374

Liu S, Tsao M (2009) Inhibition of spoilage yeasts in cheese by killer yeast *Williopsis saturnus var. saturnus*. Int J Food Microbiol 131:280–282

Liu X, Fang W, Liu L, Yu T, Lou B, Zheng X (2010) Biological control of postharvest sour rot of citrus by two antagonistic yeasts. Lett Appl Microbiol 51:30–35

Long CA, Wu Z, Deng BX (2005) Biological control of *Penicillium italicum* of citrus and *Botrytis cinerea* of grape by strain 34–9 of *Kloeckera apiculata*. Eur Food Res Technol 221(1–2):197–201

Long CA, Deng BX, Deng XX (2006) Pilot testing of *Kloeckera apiculata* for the biological control of postharvest diseases of citrus. Ann Microbiol 56:13–17

López-García B, Veyrat A, Pérez-Payá E (2003) Comparison of the activity of antifungal hexapeptides and the fungicides thiabendazole and imazalil against postharvest fungal pathogens. Int J Food Microbiol 89:163–170

Lowes K, Shearman C, Payne J, Mackenzie D, Archer D, Merry R, Gasson M (2000) Prevention of yeast spoilage in feed and food by the yeast mycocin HMK. Appl Environ Microbiol 66:1066–1076

Luo Y, Zeng K, Ming J (2012) Control of blue and green mold decay of citrus fruit by *Pichia membranefaciens* and induction of defense responses. Sci Hortic (Amsterdam) 135:120–127

Magliani W, Conti S, Travassos LR, Polonelli L (2008) From yeast killer toxins to antibodies and beyond. FEMS Microbiol Lett 288:1–8

Makovitzki A, Viterbo A, Brotman Y, Chet I, Shai Y (2007) Inhibition of fungal and bacterial plant pathogens in vitro and in planta with ultrashort cationic lipopeptides. Appl Environ Microbiol 73(20):6629–6636

Makower M, Bevan EA (1963) The inheritance of the killer character in yeast (*Saccharomyces cerevisiae*). In: Proceedings of the 11th International Congress on Genetics, vol I. Pergamon Press, Oxford, pp 202–203

Marquina D, Santos A, Peinado JM (2002) Biology of killer yeasts. Int Microbiol 5:65–71

McGuire RG (2000) Population dynamics of postharvest decay antagonists growing epiphytically and within wounds on grapefruit. Phytopathology 90:1217–1223

McGuire R, Hagenmaier R (1995) Storage waxes that support growth of *Candida oleophila* for biocontrol of *Penicillium digitatum* on citrus. Phytopathology 85:1166

Meier GE (2005) Sales de sodio como alternativas a los fungicidas tradicionales para el control de moho verde y moho azul. Actas del II Seminario Internacional de Postcosecha de Cítricos. Vázquez DE, Meier GE, Cocco M. Ediciones INTA, pp 108–110

Mercier I, Smilanick JL (2005) Control of green mold and sour rot of stored lemon by biofumigation with *Muscodor albus*. Biol Control 32:401–407

Muller I (2005) Investigaciones en poscosecha en INIA Salto Grande. Evaluación de nuevos fungicidas. Actas del II Seminario Internacional de Postcosecha de Cítricos. Ediciones INTA, Argentina, pp 84–88

Palou L (2007) Evaluación de alternativas para el tratamiento antifúngico en poscosecha de cítricos de Producción Integrada. Rev Hortic 82:93

Palou L, Smilanick JL, Droby S (2008) Alternatives to conventional fungicides for the control of citrus postharvest green and blue molds. Stewart Postharvest Rev 4:1–16

Perez MF, Contreras L, Garnica NM, Fernández-Zenoff MV, Farías ME, Sepulveda M, Dib JR (2016) Native killer yeasts as biocontrol agents of postharvest fungal diseases in lemons. PLoS One 11(10):e0165590

Perez MF, Perez Ibarreche J, Isas AS, Sepulveda M, Ramallo J, Dib JR (2017) Antagonistic yeasts for the biological control of *Penicillium digitatum* on lemons stored under export conditions. Biol Control 115:135–140

Platania C, Restuccia C, Muccilli S, Cirvilleri G (2012) Efficacy of killer yeasts in the biological control of *Penicillium digitatum* on Tarocco orange fruits (Citrus sinensis). Food Microbiol 30:219–225

Ragone ME (1999) Niveles de contaminación fúngica en galpones de empaque de exportación de frutas cítricas de la región de Concordia. Trabajo final de Graduación. Facultad de Ciencias Agrarias. Universidad Nacional del Nordeste. Corrientes, Argentina

Ren X, Kong Q, Wang H, Yu T, Zhou W, Zheng X (2011) Biocontrol of fungal decay of citrus fruit by *Pichia pastoris* recombinant strains expressing cecropin A. Food Chem 131:796–801

Rosa MM, Tauk-Tornisielo SM, Rampazzo PE, Ceccato-Antonini SR (2010) Evaluation of the biological control by the yeast *Torulaspora globosa* against *Colletotrichum sublineolum* in sorghum. World J Microbiol Biotechnol 26:1491–1502

Sánchez-Torres P, Tuset JJ (2011) Molecular insights into fungicide resistance in sensitive and resistant *Penicillium digitatum* strains infecting citrus. Postharvest Biol Technol 59:159–165

Sharma R, Singh D, Singh R (2009) Biological control of postharvest diseases of fruits and vegetables by microbial antagonists: a review. Biol Control 50:205–221

Smilanick JL, Mansour MF, Gabler FM, Sorenson D (2008) Control of citrus postharvest green mold and sour rot by potassium sorbate combined with heat and fungicides. Postharvest Biol Technol 47(2):226–223

Sulo P, Michalcakova S (1992) The K3 type killer strains of genus *Saccharomyces* for wine production. Folia Microbiol 37:289–294

Sulo P, Michalcakova S, Reiser V (1992) Construction and properties of K1 type killer wine yeasts. Biotechnol Lett 14:55–60

Suprapta DN, Arai K, Iwai H (1997) Effects of volatile compounds on arthrospore germination and mycelial growth of *Geotrichum candidum* citrus race. Mycoscience 38:31–35

Talibi I, Boubaker H, Boudyach EH, Ait Ben Aoumar A (2014) Alternative methods for the control of postharvest citrus diseases. J Appl Microbiol 117(1):1–17

Taqarort N, Echairi A, Chaussod R, Nouaim R, Boubaker H, Benaoumar AA, Boudyach E (2008) Screening and identification of epiphytic yeasts with potential for biological control of green mold of citrus fruits. World J Microbiol Biotechnol 24:3031–3038

Torres Leal GJ, Velázquez PD, Paz AE, Farías MF (2008) Control of Penicillium digitatum (Green mold) by sodium bicarbonate in lemon fruit in Tucuman (Argentina). In: Proceedings of the international society of citriculture, XI Congress, Wuhan, China, p 1369

Tripathi P, Dubey N (2003) Exploitation of natural products as an alternative strategy to control postharvest fungal rotting of fruit and vegetables. Postharvest Biol Technol 32:235–245

Tuset JA (1987) Podredumbres de los frutos cítricos. Generalitat Valenciana, Valencia, p 206

Utama IMS, Wills RB, Ben-yehoshua S, Kuek C (2002) In vitro efficacy of plant volatiles for inhibiting the growth of fruit and vegetable decay microorganisms. J Agric Food Chem 50(22):6371–6377

Vázquez DE, Ragone M, Garrán S (1995) Factores que afectan la calidad de los frutos cítricos. Informe técnico de la Estación Experimental Agropecuaria Concordia del INTA, p 14

Velázquez PD, Farías MF, Carbajo MS, Torres Leal GJ (2010) Eficacia de los fungicidas azoxistrobin fluodioxinil en el control curativo de "moho verde" causado por *Penicillium digitatum* en frutos de limón. Libro de Resúmenes XXXIII Congreso Argentino de Horticultura, Rosario, p 142

Viñas I (1990) Principios básicos de la patología de poscosecha. FRUT 5:285–292

Wang YF, Bao YH, Shen DH, Feng W (2008) Biocontrol of *Alternaria alternata* on cherry tomato fruit by use of marine yeast *Rhodosporidium paludigenum* Fell & Tallman. Int J Food Microbiol 123:234–239

Wilson CL, Chalutz E (1989) Postharvest biological control of *Penicillium* rots with antagonistic yeasts and bacteria. Sci Hortic 40:105–112

Wisnieswski ME, Biles C, Droby S, McLaughlin R, Wilson C, Chalutz E (1991) Mode of action of the postharvest biocontrol yeast, *Pichia guilliermondii*. Characterization of attachment to *Botrytis cinerea*. Physiol Mol Plant Pathol 39:245–258

Wisnieswski M, Wilson C, Droby S, Chalutz E, ElGhaouth A, Stevens C (2007) Postharvest biocontrol: new concepts and applications. In: biological control: a global perspective. CAB International, Wallingford, pp 262–273

Yao HJ, Tian SP (2005) Effects of a biocontrol agent and methyl jasmonate on postharvest diseases of peach fruit and the possible mechanisms involved. J Appl Microbiol 98:941–950

Zhang HY, Zheng XD, Xi YF (2005) Biological control of postharvest blue mold of oranges by *Cryptococcus laurentii* (Kufferath) Skinner. BioControl 50(2):331–342

Zhang D, Spadaro D, Garibaldi A, Gullino ML (2010) Selection and evaluation of new antagonists for their efficacy against postharvest brown rot of peaches. Postharvest Biol Technol 55:174–181

Zheng XD, Zhang HY, Sun P (2005) Biological control of postharvest green mold decay of oranges by *Rhodotorula glutinis*. Eur Food Res Technol 220:353–357

Chapter 8
Disease-Suppressive Effect of Compost Tea Against Phytopathogens in Sustaining Herbal Plant Productivity

Abd. Rahman Jabir Mohd. Din, Siti Zulaiha Hanapi, Siti Hajar Mat Sarip, and Mohamad Roji Sarmidi

Abstract As for all other important crops, soilborne diseases in herbal plants are becoming a complex and quite prevalent problem that significantly reduced yield production, which may severely influence the bioactive compounds. These valuable crops are taking an increasing relevance in mass, and the dependence on chemical fungicides should be reduced to preserve the qualitative aspect of production. One of the available alternative biological approaches is the compost teas. It is strongly imperative to use it to prevent, suppress, or control wide range of soilborne plant diseases especially caused by fungal phytopathogens. Compost teas are a fermented compost aqueous extract that consists of abundance of antagonistic microbes within it. This review provides the key principles to several aspects of inhibitory potential of compost teas derived from biological origins including fundamental idea for the preparation of compost tea, bio-efficacy effects in managing the plant diseases, as well as mechanism on disease suppression. Better understanding of the antagonism microbial interactions within compost teas will determine the sustainability of any approaches for a biocontrol program. The factors affecting their effectiveness also will be briefly explained. Finally, future research is proposed to further validate the strategies of compost teas as suitable tool of plant protectants toward an improvement of the overall health of plants.

A.R.J. Mohd. Din (✉)
Innovation Centre in Agritechnology for Advanced Bioprocess (ICA),
UTM Pagoh Research Center, Pagoh Education Hub, 84600, Pagoh, Johor, Malaysia
e-mail: jabir@ibd.utm.my

S.Z. Hanapi • S.H.M. Sarip
Faculty of Chemical and Energy Engineering, Institute of Bioproduct Development (IBD),
UTM, 81310, Johor Bahru, Johor, Malaysia

M.R. Sarmidi
Innovation Centre in Agritechnology for Advanced Bioprocess (ICA),
UTM Pagoh Research Center, Pagoh Education Hub, 84600, Pagoh, Johor, Malaysia

Faculty of Chemical and Energy Engineering, Institute of Bioproduct Development (IBD),
UTM, 81310, Johor Bahru, Johor, Malaysia

Z.A. Zakaria (ed.), *Sustainable Technologies for the Management of Agricultural Wastes*, Applied Environmental Science and Engineering for a Sustainable Future, https://doi.org/10.1007/978-981-10-5062-6_8

8.1 Introduction

In Malaysia, abundance of herb diversity (more than 120 species, from shrubs to large trees) has created huge opportunity for the development of various industries, notably the wellness industry, as these herbs are perceived medically safe (Abas et al. 2006). However, the control of fungal disease on these herbs requires a high level of attention. Plant protection played an extremely crucial role in improving productivity rate toward preservation of therapeutic claim of efficacious high-end herbal products. Compost teas have received increased attention over the last few years for plant disease management. In spite of this, modern agricultural practice still relies heavily on the application of synthetic agrochemicals for the control of various plant disease-causing phytopathogens. Sustainable and environmental-based biological control approaches need to be translated by reducing the need of hazardous agrochemicals to maintain plant health and productivity. On the other hand, the use of biological control agents such as *Trichoderma*, *Bacillus*, and *Streptomyces* in experimental trials resulted in inconsistencies in plant disease control. In response to cater public health concern for sustainable production, the reason of all these "failures" was related to inability of single organism to work properly with constraint of numerous biotic and abiotic factors. Compost tea offers an interesting alternative as plant disease control agent based on its high microbial diversity content especially beneficial microorganisms which is a key factor for inhibiting the development of plant disease through their suppressive antagonistic abilities (Haggag and Saber 2007; Yohalem et al. 1996; Kone et al. 2010; Siddiqui et al. 2011). St. Martin et al. (2012) reported that suppressive properties in compost tea consist of the diverse modes of antagonist microbes that help to survive feasibly in extremely harsh environment. Generally, non-aerated compost tea (NCT) showed greater suppression on phytopathogens compared to aerated compost tea (ACT). Efficacy assessment of compost teas was mentioned here as a statistically significant reduction in plant disease severity comparative to water-treated control over host plant tested through many approaches such as detached leaf assays, glasshouse, and field trials. Compost teas also could be easily introduced into existing disease management control program used in conventional agriculture.

8.1.1 Sustainable Biological Management of Herbal Plant Diseases

Agricultural sustainability in herbal plant production is greatly depending on the development of strategies to reduce the high reliance on the utilization of agrochemicals. Herbs are similar to other horticultural crops as they also are susceptible to a variety of diseases that resulted in significant production problems that influence both yield and overall finished product quality. Besides the obvious environmental-related problems, continuous and uncontrolled application of agrochemicals may also result in the development of gene-resistance genotype and serious devastation on existing plant and its plantlets (Janisiewicz and Korsten 2002). Soilborne phytopathogens have become major threat to farming practice. Many factors contribute to the increase in severity of soilborne diseases such as inappropriate irrigation management, lack of crop rotation practice, and use of disease-susceptible genotype (Siddiqui et al. 2011). Soilborne fungal phytopathogens produce small propagules that inhibit microbiostasis which in turn activated nutrient competition and antibiosis. Among the phytopathogens, *Alternaria* spp. and *Fusarium* spp. are two peculiar fungi that cause catastrophic symptoms such as vascular wilt, damping-off, or even worse whitish powder covering foliage. Devastating outbreaks can lead to leaf necrosis and chlorosis, defoliation, stem lesion, and stunted growth followed by serious multiple phytopathogenic infections (Litterick et al. 2004).

Other practices commonly used in managing soilborne phytopathogens include biofumigation, soil disinfestation, and soil solarization (Gimsing and Kirkegaard 2009; Messiha et al. 2007; Katan 2015). Plant disease management program nowadays is aimed at reducing or eliminating the use of synthetic agrochemicals by integrating the biological control inputs. This technique is relevant as long as they have the potential to suppress a broad range of phytopathogens. Controlling the soilborne diseases cannot be measured effectively if a single or conventional management strategy was applied. One of the biocontrol agents that is viewed as potential alternatives to chemical fungicide is compost teas. This natural biocontrol agent played a promising role in response to the increasing demand of environmental sustainability for farming practices and herbal product safety. In the past decades, compost teas have been scientifically proven to control the broad range of phytopathogens responsible for wilting, melanose, lesion, and damping-off in many cropping systems including *Fusarium* spp., *Pythium* spp., *Sclerotinia* spp., *Verticillium dahliae*, *Phytophthora* spp., *Rhizoctonia solani*, *Podosphaera fusca*, and *Lecanicillium* spp. (Marin et al. 2013).

One of the current development strategies in the application of diverse biocontrol agents into herbal farming is the integration of antagonistic microbes into microbial consortia that play a key role in disease-suppressive mechanism. Inoculation of these antagonist properties into several carrier materials such as compost, peat, and alginate compounds was suitable for effective biocontrol against pre- and postemergence of damping-off disease severity. Esfahani and Monazzah (2011) described the prevalence of fungal diseases in medicine herbal plants including rosemary, lavender, sage, viper's bugloss, burdock, and pumpkin seeds. Chae et al. (2006) suggested the compost fortified by consortium of antagonistic microbes could act as soil amendment for the control of disease caused by late blight (*Phytophthora capsici L.*) in pepper. Lagerlof et al. (2011) evaluated the effectiveness of fungivorous nematodes and compost with disease-suppressive compost to inhibit damping-off caused by *Rhizoctonia solani* in seedlings. Outbreak of *Carica* papaya seedlings foot rot caused by *Pythium aphanidermatum* was successfully controlled by adding the specifically formulated arbuscular mycorrhizal fungi in biofertilizer (Olawuyi et al. 2014). This also was in consonance with Romero et al.'s (2004) findings that antagonistic microbial inoculants (*Bacillus* sp.) proved to be efficacious in control of cucurbit powdery mildew. According to Kalra et al. (2003), the use of induction of systemic resistance in host plant combined with fungicide application was successful for leaf blight control in menthol mint (*Mentha arvensis*). Xu et al. (2012) found compost teas to be effective against soilborne pathogenic fungi in cress germinated seeds, types of biennial herbs. Recently, Mohd Din et al. (2016) evaluated the suppressive effect of compost teas as a foliar spray to challenge the leaf spot symptom in the local edible herb (*Melicope ptelefolia*) leaves.

Akila et al. (2011) reported the efficacy of combination of botanical extracts and antagonistic inoculants (*Pseudomonas fluorescens* 1 and *Bacillus subtilis*) in reducing the *Fusarium* wilt severity for both greenhouse and field test assay in banana cultivation. This kind of a novel consortium showed the significant reduction in the disease outbreak and enhanced sustainable growth of crop promotion. Gava and Pinto (2016) recently successfully proved the synergistic effect of compatible mixture of *Trichoderma polysporum* LCB 50 strain and compost teas to control melon wilt caused by *Fusarium oxysporum* f. sp. *melonis* (Fom) under field conditions. These strategies are quite formidable and could act as a launchpad toward sustainable herbal plantation utilizing biological control management program.

8.2 Production Strategy and Application of Compost Tea

8.2.1 Aerated and Non-aerated Compost Tea

Many terminologies have been used to refer to compost teas, and all of them share the same meaning. Some definitions are used interchangeably and always referred to the compost tea production, namely, compost extract (Weltzien 1991) and compost liquid (Gava and Pinto 2016). All these terms refer to aqueous samples obtained from composted materials by fermentation within certain period of incubation time with or without active aeration. Compost teas can be further classified into aerated and non-aerated compost teas with regard to the fermentation and brewing methods used. Scheuerell and Mahaffee (2002) suggested two dominant methods for the production of compost teas. Aerated compost teas (ACTs) are produced using a customized container containing continuous aeration of dechlorinated water to proliferate the microbial activity and take place between 24 and 72 h. Whereas, the non-aerated compost teas (NCTs) involved minimal or no aeration within the 7–14-day contact period (Litterick and Wood 2009). Both ACT and NCT production methods involved brewing the compost in water for a specific time period and required the use of a brewing container, microbial additive, composted materials, dechlorinated water, incubation, and filtration prior to application. Naidu et al. (2010) stated that addition of microbial additives at the beginning or during the fermentation process of compost teas was carried out to stimulate the beneficial microbial populations by providing nutrients such as molasses, fish hydrolysate, rock dust, or humic acid. Yohalem et al. (1994) suggested 4 months of storage period (at 20 °C) as the optimal time to avoid any loss of suppressiveness potential against phytopathogens.

There are various devices designed by commercial companies to produce ACTs. One example is suspending porous compost bag actively bubbled up by aquarium stones to create recirculating water all the way. This can be done in the open container where water will be recirculated through a vortex nozzle mounted above a tank and air injection through a hollow propeller shaft or fine bubble diffusion mats (Ingham and Alms 1999; Merrill and Mckeon 2001). Shrestha et al. (2011a, b) recommended a compost-to-water ratio of 1:10 (v/v) to maximize the physicochemical and microbial properties of compost teas. At present, the best compost-to-water ratio for ACT and NCT brewing is 1:2.5 (v/v) with 2 days of fermentation time as suggested by Islam et al. (2016). All compost teas need to be stirred every 2–3 days to ease the liberation of microbes from porous compost bag (Brinton and Droffner 1995). The amount of compost used is totally dependent on the size and type of the brewing container and equipment used.

Siddiqui et al. (2009) prepared ACT and brewed it for 12 days prior to its use as a disease suppressor for crop pathogen in okra. Moreover, the incubation period of compost tea prepared by Kone et al. (2010) required 14 days, whereas Naidu et al. (2010) recommend 7 days of incubation. The present study was similar to Naidu et al. (2010) that sufficient incubation length is needed to promote plant growth. NCT was better compared to ACT as suggested by Scheuerell and Mahaffee (2006). Shorter brewing time, lower or no phytotoxicity effect, and increased microbial mass are some of the properties that are usually linked with ACTs (Scheuerell and Mahaffee 2002). Ingram and Millner (2007) advocated that the capacity of human pathogenic bacteria regrowth relied on the type of nutrient additives at the commencement stage of the brewing process. Duffy et al. (2004) reported that addition of molasses supplement can stimulate pathogen regrowth (*Salmonella* spp. and *E. coli* O157:H7) in ACTs tested and dependent on the type of composted materials used. Kannangara et al. (2006) reported that anaerobic condition during the production of non-aerated compost teas could lead to the possibility of *E. coli* regrowth.

Most NCTs showed greater suppression on phytopathogens compared to ACTs. This also comes with varied results and findings. Ingham and Alms (1999) claimed that ACT tends to have suppressive ability because of higher microbial population, and availability of nutrient additive helps to proliferate number of beneficial microorganism for inhibitory agent, *Botrytis cinerea*. There are few reports on the disease-suppressive effect of compost tea over soilborne and airborne phytopathogens (Dionne et al. 2012; Sang et al. 2010; St. Martin et al. 2012). Indeed, Marin et al. (2013) have made a comprehensive study regarding compost teas from different sources against eight phytopathogens.

Compost teas can be applied either as foliar spray or drenched in soil to suppress a wide range of diseases caused by phytopathogens (Scheuerell and Mahaffee 2002; Al-Mughrabi et al. 2008; Kone et al. 2010; Dukare et al. 2011; Naidu et al. 2013). Islam et al. (2014) evaluated the suppressive effect of compost tea applied as soil drench to activate plant defense pathway of host plants against disease outbreak in brinjal. Foliar spray of compost teas on leaf surfaces typically changes the set of phyllosphere on foliage and induces resistance in host plants for short periods of time. Siddiqui et al. (2009) had successfully used aerated compost teas as foliar spray against pathogen of wet rot in okra and suggested that plants should be sprayed at short intervals to retain the inhibitory effect.

8.3 Physicochemical Analysis of Compost Teas

Several factors influenced the physicochemical components of compost teas which are responsible for their efficacy in plant disease suppression involved in compost tea extraction process. Nonetheless, disease-suppressive effects have been attributed to organic and inorganic properties present or released by microorganisms inhabiting these inputs. Most important thing, physicochemical properties greatly depend on the compost maturity quality itself. Humic acid, phenolics, soluble mineral nutrients, bioactive compounds, plant growth regulators, and volatile fatty acids have been suggested as organic properties which play an important role in disease-suppressive effects (Siddiqui et al. 2008; Pant et al. 2012). Unlike other reports, Zmora-Nahum et al. (2008) observed that the dominant inhibitory properties come from heat-stable and nonprotein metabolites released under anaerobic environment particularly from non-aerated compost teas. Pane et al. (2016) reviewed extensively on compost tea bioactivity in particular dissolved and metabolite substances including humic moieties and little organic molecules that lead to malformed mycelium morphology of fungal phytopathogen. Most of compost teas contained high level of N (total N, NO_3-N, NH_4-N), and the increase level of this important nutrient is greatly affected by compost/water ratio, extraction time, and storage duration as described by Islam et al. (2016). However, chemical properties of compost teas especially total N and pH did change significantly from the storage period with the exception of microbial population. Release of N is triggered through the different biochemical metabolic pathways such as mineralization, transformation, and stabilization of organic composted materials through the formation of humic acid substances (Senesi et al. 2007; Said-Pullicino et al. 2007). Furthermore, Borrero et al. (2012) mentioned that high level of N and NH_4:N ratios have contributed to reduce the *Fusarium* wilt disease severity in tomato plants.

Xu et al. (2015) reported on the physicochemical analysis of pig slurry compost teas under the influence of different aeration rate quantities by combining the FTIR, CP-MAS, ^{13}C NMR, and EEM methods. All techniques were used to elucidate the detailed structure characterization of compost teas. EEM spectra verified the presence of humic-like substances as these compounds were most likely to be the dominant components in all compost teas tested. The total water-soluble organic carbon, total N, total P, and humification degree were gradually increased as the aeration rate elevated, where 11 L min^{-1} was concluded as the most feasible aeration rate for the production of good compost tea. In addition, Carballo et al. (2008) investigated the chemical composition of compost teas using FTIR spectroscopy and thermal analysis. From the FTIR spectra obtained, compost teas were shown to contain peaks that were attributed to the presence of primary and secondary amine groups. Sang et al. (2010) suggested that the nature of disease-suppressive effects of compost teas was likely due to activity of heat-stable chemical compounds.

Meanwhile, Pant et al. (2012) analyzed plant growth regulators such as cytokinin isopentenyladenosine (iPA) and abscisic acid from the green waste thermophilic compost teas. They detected some of the gibberellins such as gibberellin 4 (GA_4), gibberellin 24 (GA_{24}), and gibberellin 34 (GA_{34}) in the food waste vermicompost and chicken manure-based vermicompost teas, respectively. Humic substances were reported to have auxin-like activity that influences nitrate metabolism for growth improvement (Eyheraguibel et al. 2008; Muscolo and Sidari 2009). Said-Pullicino et al. (2007) found active lignin-derived phenols such as vanillyl, syringyl, cinnamyl, and hydroxyl phenols in the stable hydrophobic portion of compost teas, whereas polysaccharidic moieties were identified in the hydrophobic portions.

To this end, many researchers concluded that compost teas suppress phytopathogens through general mechanisms, rather than specific, since competition and production of antibiotics are more often reported.

8.4 Compost Tea Microbiology

Microbial populations within compost teas were cited as the main factor responsible for the inhibitory effects (Siddiqui et al. 2009; Dionne et al. 2012). Chemical and microbiological properties in compost teas have shown to work in parallel to suppress phytopathogens (Castano et al. 2011). Suppression effect of compost teas is greatly depended on the antagonistic microbe colonization residing within the compost teas. Different groups of microbial populations residing within compost teas can be classified into facultative plant symbionts and competitive saprophytes (Mehta et al. 2014). Some of the cultures identified to possess the same function in disease suppressiveness include the genera of *Bacillus*, *Penicillium*, and *Trichoderma* and those proteolytic bacteria such as *Enterobacteria*, *Serratia*, *Nitrobacter*, *Pseudomonads*, and *Actinomycetes* (Naidu et al. 2010; Marin et al. 2013). All these microbes exerted the ability to colonize and express biocontrol mechanism on phyllosphere (plant surface) and rhizosphere. *Trichoderma*, a mycoparasitic fungus, was found to act as a primary antagonist in suppressing the disease incidence. The maturity of compost was determined after undergoing three essential different phases, namely, mesophilic phase (up to 40 °C), thermophilic phase (over 40 °C), and maturation phase (up to 40 °C). All these phases constituted different concentration of microbial population present in the final product of composted materials responsible to disease suppression effect (Raj and Antil 2011).

Zhou et al. (2016) highlighted the presence of a total of 17 bacterial and 22 fungal species during the maturation phase for Chinese medicine herbal residual compost. Table 8.1 described the variation of microbiological properties among the tested compost teas.

Table 8.1 Population of various microbes in different types of compost teas

Type of compost	Microbiological properties, CFU ml^{-1} @ CFU g^{-1}					References
	Bacteria	Fungi	Yeast	*Pseudomonas* spp.	*Actinomycetes*	
Compost tea incubated at 20 °C	6.7	2.5			2.1	Hegazy et al. (2015)
Compost tea incubated at 28 °C	7.1	2.5	NA	NA	2.3	
Compost tea incubated at 37 °C	6.7	3.1			2.4	
Compost tea incubated at 45 °C	6.2	3.5			2.6	
Whey compost	9.0×10^{3}	3.51×10^{2}	2.1×10^{5}	2.0×10^{3}		Pane et al. (2013)
Water compost	8.3×10^{3}	2.81×10^{2}	2.1×10^{5}	2.0×10^{3}	NA	
Chicken manure vermicompost (aged)	6.9	0.3				Pant et al. (2012)
Chicken manure thermophilic compost	6.7	0.2				
Food waste vermicompost	6.7	0.4	NA	NA	NA	
Chicken manure vermicompost (fresh)	6.7	0.4				
Green waste thermophilic compost	6.8	0.1				
Compost + yeast extract	2.1×10^{9}	4.3×10^{6}	1.3×10^{7}	6.7×10^{8}	3.9×10^{8}	Naidu et al. (2009)
Compost + peptone	2.3×10^{9}	4.4×10^{5}	1.6×10^{5}	5.0×10^{8}	4.4×10^{7}	
Compost + brown sugar	1.5×10^{9}	6.3×10^{6}	6.3×10^{6}	3.8×10^{8}	4.2×10^{7}	
Compost + humic acid	9.6×10^{8}	1.8×10^{7}	1.8×10^{7}	4.5×10^{7}	8.0×10^{8}	
Compost + kelp	1.6×10^{9}	5.2×10^{6}	5.2×10^{6}	4.0×10^{8}	8.0×10^{7}	
Compost + corn meal	3.1×10^{8}	3.8×10^{6}	3.8×10^{6}	9.8×10^{7}	1.5×10^{8}	

(continued)

Table 8.1 (continued)

Type of compost	Microbiological properties, CFU ml^{-1} @ CFU g^{-1}					References
	Bacteria	Fungi	Yeast	*Pseudomonas* spp.	*Actinomycetes*	
Chicken manure	3.2×10^{6}	3.2×10^{3}		7.0×10^{2}	1.8×10^{4}	Kone et al. (2010)
Bovine manure	3.2×10^{6}	2.8×10^{3}		6.0×10^{1}	7.0×10^{4}	
Sheep manure	1.2×10^{7}	1.4×10^{3}	NA	1.0×10^{3}	2.3×10^{5}	
Shrimp	1.2×10^{6}	2.8×10^{3}		7.0×10^{2}	2.2×10^{4}	
Seaweed	1.0×10^{6}	3.2×10^{3}		9.0×10^{2}	3.2×10^{4}	
Organic compost	7.1×10^{7}	1.1×10^{4}	NA	NA	NA	El-Din and Hendawy (2010)
Rice straw + [a]EFB compost	9.2×10^{12}	7.9×10^{6}	NA	4.8×10^{5}	2.3×10^{4}	Siddiqui et al. (2009)
Rice ash ([a]ACT)	4.2×10^{11}	4.9×10^{4}	2.3×10^{6}		1.7×10^{5}	Haggag and Saber (2007)
Bean straw (ACT)	3.7×10^{11}	4.6×10^{4}	3.7×10^{6}		1.3×10^{5}	
Vegetative food waste (ACT)	5.0×10^{11}	7.0×10^{4}	3.2×10^{6}		2.5×10^{5}	
Chicken manure (ACT)	5.4×10^{11}	7.9×10^{4}	4.1×10^{6}	NA	3.0×10^{5}	
Rice ash + chicken manure (ACT)	6.1×10^{11}	9.9×10^{4}	4.0×10^{6}		3.8×10^{5}	
Bean straw + chicken manure (ACT)	5.6×10^{11}	6.5×10^{4}	5.5×10^{6}		3.3×10^{5}	
Vegetative food waste + chicken manure (ACT)	6.9×10^{11}	9.9×10^{4}	4.3×10^{6}		3.8×10^{5}	
Horse-chicken manure	5.6×10^{10}	1.1×10^{2}	NA	NA	2.4×10^{5}	McQuilken et al. (1994)

[a]*EFB* empty fruit bunch of oil palm, *ACTs* aerated compost teas

The complexity and dynamism of microbial population in compost teas need to be understood in order to improve the efficacy of disease suppression for different varieties of phytopathogens. A number of researchers have revisited the potential use of molecular assessment to ascertain the microbial diversity complexity in compost teas such as polymerase chain reaction (PCR) analysis corroborated with other techniques including DNA sequencing, denaturing gradient gel electrophoresis (DGGE), and terminal restriction fragment length polymorphism (T-RFLP) (Shrestha et al. 2011a, b; Fritz et al. 2011; Mehta et al. 2014; Pane et al. 2014). Pane et al. (2013) also successfully characterized functional metabolic groups responsible for disease-suppressive effect against *Rhizoctonia solani* and *Sclerotinia minor* using the Biolog system. Alternatively, advanced metabolomic approaches could be possibly used to investigate the role of metabolites in disease suppression and plant defense responses.

8.5 Assessment and Efficacy of Compost Tea in Managing Plant Diseases

Compost teas reported to possess both protective and nutritive effects on herbal plants. Most of the scientific studies demonstrated the potential of compost teas as crop protectants with in vitro screening activity against phytopathogens. Many reports evaluated the compost teas as a biological control over phytopathogens in the field trial scale although many claimed it to be too costly and risky for environmental condition. Siddiqui et al. (2011) and Javanmardi and Ghorbani (2012) stated the efficacy of compost teas on the morphological effect on *Centella asiatica* and basil, respectively. Previously, El-Din and Hendawy (2010) discovered that compost tea application along with dry yeast could significantly enhance oil composition of *Borago officinalis* herbal plant. However, there are limited studies indicating the biocontrol effect of compost teas against phytopathogens in herbal plants compared to reports for horticultural plants. Table 8.2 showed the proven assessments of compost teas in herbal and non-herbal plants following the disease severity reduction.

In vitro screening assay could not be a good predictor to evaluate the compost tea efficacy as pathogenesis is directly influenced by different field conditions. The use of in vitro screening only could lead to misinterpretation on the complexity of microbial populations and interaction with pathogen. The effect of biotic and abiotic factors must also be taken into consideration, as failure to include these factors would result in inaccurate final mass production result. One of the common failures of compost teas is to express the suppressiveness ability against aerial plant diseases which can be attributed to the dynamic nature of field environment. The fundamental basis for the efficiency of compost tea depends on its ability to change the microbiota population in the soil rhizosphere or plant growth substrate as a whole. Most recent studies nowadays have focused on the compost tea efficacy across the types and other biocontrol agents instead of comparing with the commercial synthetic fungicides because this will directly answer the practical purpose toward achieving organic crop production (Kabir et al. 2012; Albert et al. 2012; Pane et al. 2013; Naidu et al. 2013). Pane et al. (2013) found that the presence of *Botrytis cinerea*, *Alternaria alternata*, and *Pyrenochaeta lycopersici* in tomato was reduced in plants sprayed with whey compost teas compared with those sprayed with conventional water compost teas.

Table 8.2 Types of phytopathogens and plant treated by compost teas produced from herbal and non-herbal based

Composted materials	Plant	Phytopathogens	Brewing method	References
Non-herbal plant				
Cow, sheep, horse, and chicken manures and *Trichoderma* spp.	Tomato	*Fusarium oxysporum* f. sp. *lycopersici*	ACT[a]	Moosa et al. (2016)
Vermicompost	Bean	*Tetranychus urticae*	ACT	Aghamohammadi et al. (2016)
Compost poultry litter	Brinjal	*Ralstonia solanacearum*	ACT	Islam et al. (2014)
Agro-waste compost	Rose	*Sphaerotheca pannosa*	ACT	Seddigh et al. (2014)
Chicken manure and timber residual compost	Grape vine	*Botrytis cinerea*	ACT	Evans et al. (2013)
Biowaste compost and composted tomato residues	Tomato	*Alternaria alternata*, *Botrytis cinerea*, and *Pyrenochaeta lycopersici*	ACT	Pane et al. (2013)
Chicken manure, sheep manure, bovine manure, shrimp powder, or seaweed	Tomato	*Alternaria solani*, *Botrytis cinerea*, and *Phytophthora*	NCT	Kone et al. (2010)
Farmyard manure, poultry manure, vermicompost, spent mushroom compost, *Lantana camara*, and *Urtica* sp.	French bean	*Rhizoctonia* root rot (*Rhizoctonia solani* Kühn) and angular leaf spot (*Phaeoisariopsis griseola* (*Sacc.*) *Ferraris*)	ACT	Joshi et al. (2009)
Agro-waste such as rice straw (RST) and empty fruit bunch (EFB) of oil palm composts	Muskmelon	*Choanephora cucurbitarum*	ACT	Siddiqui et al. (2009)
Market and garden wastes	Tomato	*Erysiphe polygoni*	ACT	Segarra et al. (2009)
Vermicompost	Rice	*Fusarium moniliforme*	ACT	Manandhar and Yami (2008)
Plant residues (rice ash, bean straw, and vegetative food waste) and chicken manure	Onion and tomato	*Alternaria solani* and *Alternaria porri*	NCT and ACT[a]	Haggag and Saber (2007)

(continued)

Table 8.2 (continued)

Composted materials	Plant	Phytopathogens	Brewing method	References
Thermal compost, static wood chips, and vermicast	Potato tubers	*Phytophthora infestans*	ACT	Al-Mughrabi et al. (2008)
Herbal plant				
Agro-waste compost	*Melicope ptelefolia* (tenggek burung)	*Gammothele lineata*	NCT	Mohd Din et al. (2016)
Khaya and *Eucalyptus* composts	Basil plant	*Fusarium oxysporum*	ACT	Hassan and Abo-Elyousr (2013)
Pig manure, cow manure, sawdust, and zeolite	Pepper and cucumber	*Colletotrichum coccodes* and *C. orbiculare*	ACT	Sang and Kim (2011)
Commercial compost from six facilities in Korea	Pepper	*Phytophthora capsici*	ACT	Sang et al. (2010)
Vegetable waste compost	Cucumber	*Cucumber mosaic virus*	NCT	Wahyuni et al. (2010)
Chitin compost (crab shell, vermiculite, rice straw, rice bran)	Pepper	*Phytophthora capsici*	NCT	Chae et al. (2006)

[a]*ACT* aerated compost tea, *NCT* non-aerated compost

8.5.1 In Vitro Assay

This assay is carried out to determine the in vitro inhibitor effect which recorded the inhibition zone generated by each compost teas against well-grown phytopathogenic fungi plug. This technique provided the complete profile for mycelial growth without physical interaction. Agar plugs with actively growing mycelia of phytopathogenic fungi were individually inoculated on the compost tea-amended media for 7 days in the dark at 25 °C. Percentage of inhibition in radial growth (PIRG) was determined as described by Jinantana and Sariah (1998) or Bernal-Vicente et al. (2008) according to experimental design. Results recorded that non-sterilized compost teas inhibited all eight phytopathogenic fungi regardless of extraction methods and compost/water ratio tested with exception of microfiltrated compost teas (Marin et al. 2013). Compost teas significantly suppressed phytopathogenic fungi growth on the in vitro assay, providing an early assessment for the inhibitory of soilborne plant diseases. Chinese medicine herbal residual compost tea was reported to exhibit antifungal properties against *Alternaria solani* and *Fusarium oxysporum* through in vitro assay.

8.5.2 Detached Leaf Assay

The detached leaf assay provides a rapid screening method to evaluate how pathogenesis occurred. This assay is generally conducted in vitro under highly controlled environments and beneficial to assess the resistance screening of phytopathogen-challenged host plants. However, this technique generally will be further validated in a greenhouse trial. A spore suspension adjusted to 10^5 spores ml^{-1} of phytopathogenic fungi was sprayed over the leaf to form the droplets. Palmer et al. (2010) reported the suppression of *Botrytis cinerea* severity on the compost tea-treated *Vicia faba* leaf. Disease severity scale was assessed according to Santos et al. (2008) and further evaluated based on formula described by Mitra et al. (2013).

8.5.3 Greenhouse and Field Trial

Marin et al. (2013) evaluated the effectiveness of compost tea from grape marc on improving the severity of melon disease. Plants were challenged with gummy stem blight and powdery mildew suspension at the two-leaf stage. Both aerated and non-aerated vermicompost teas proved to be effective after only 1.4% total leaves were severely infected. Haggag and Saber (2007) reported that a higher yield of tomato and onion was obtained after the application of compost teas on *Alternaria* blight disease at the field experiment. Soil drench experiment of compost tea was demonstrated by Xu et al. (2012) to assess the severity of root-knot nematode in tomato pot experiment in a greenhouse. They revealed that application of ACTs had significantly increased the root fresh biomass of tomato, thus reducing the disease incidence by 78%. Pane et al. (2013) also demonstrated significant suppressive effects of whey compost teas against *Botrytis cinerea* and *Alternaria alternata* in vivo trial of tomato plants taking into consideration suitable subsequent dilution on foliage to avoid any lesions due to phytotoxicity symptoms.

8.6 Disease Suppression Mechanism of Compost Tea

Fungal phytopathogens can be managed through one or more biological disease suppression mechanisms. Total microbial biomass and physiological profiling structure of microbial communities have been associated with multiple disease suppression of various diseases (Nobel and Coventry 2005). The presence of beneficial microorganisms may act as pathogen antagonists based on their ability to compete for nutrients or infection sites (Al-Mughrabi et al. 2008), parasitism (El-Masry et al. 2002), and activation of disease-resistance genes through induced systemic resistance in host plants (Zhang et al. 1998). In addition, secretion of secondary metabolites on the plant surface with the application of compost teas has shown to be one of the mechanisms underlying antimicrobial activity (Sang and Kim 2011). Chemicals produced by microbes such as siderophores, pseudobactins, surfactins, and other antifungal compounds were likely to be the reason for greater inhibition as these could develop nutrient sink by creating an iron starvation condition for the phytopathogens (On et al. 2015). Dianez et al. (2006) specifically mentioned the important role of siderophores which capture the iron bioavailability from being taken by fungal pathogen, thus easily facilitating the death of the particular plant diseases. These microbial siderophores could act as a potential defense mechanism, i.e., a potential biocontrol agent. Under iron-limiting environment, the soilborne causal pathogens reduced its own capability to induce pathogenesis toward the host due to the iron sequestration that occurred (Saha et al. 2016).

El-Masry et al. (2002) described on how antagonistic process against phytopathogenic fungi occurred with corroboration of the hydrolytic activity produced by those microbes. The presence of threshold density of microbial population is key to the success in biocontrol approach. The production of hydrolytic enzymes which were also known as fungal cell wall-degrading enzymes (cellulose, protease, and chitinase) has been proven to be reported as a mechanism of soilborne phytopathogen suppression (Haran et al. 1996; Siddiqui et al. 2008). Once the antagonist microbes confronted the phytopathogenic fungi host, its hyphae would coil to form the hook-like structures before penetration occurred to degrade its cell wall. Scheuerell and Mahaffee (2004) suggested a microbial population of 10^6 CFU mL^{-1} as the threshold value for the transition from non-suppressive to suppressive mechanism.

Another mechanism known as antibiosis was further clarified by Elad and Freeman (2002) as the penetration of toxic metabolites into the cell of phytopathogens that caused lysis and chemical toxicity. Antibiosis, also known as antibiotic-mediated suppression, signifies microbes that secrete one or more compounds linked to lethal activity for broad range of phytopathogens. For example, Hariprasad et al. (2011) reported the presence of chitinolytic enzymes secreted by rhizobacteria within compost which were implied to control against severity of *Fusarium* wilt in tomato crops. Siddiqui et al. (2009) also recorded that the vital suppression mechanism by compost teas was a result in the mycelia lysis of phytopathogens. Ruptured *Golovinomyces cichoracearum* conidia were reported since it had been treated with microbial-enriched compost teas which could be explained through leakage of cellular wall boundary.

8.7 Conclusion and Future Outlook

Successful application of compost teas for controlling plant diseases necessarily is based on specific cultivars and specific fungal phytopathogenic species. However, this rising attention on compost teas as disease suppressor used as foliar spray must be followed by the standardization quality, overcoming the inconsistency produced result among the bulk of related research publications. Failure of biological control mostly comes from lack of ecological fitness of the microbial communities residing within compost teas. This means sustaining the acceptable threshold level needed to be effective. Furthermore, all findings need the specific evaluation in terms of the different modes of compost tea preparation to treat different diseases under natural stressor biotic factors. This includes mode of pathogenicity of plant pathogens and different responses of compost teas, respectively. Target-specific biocontrol agents and right specific timing generally needed to control plant diseases in particular climates.

Moving forward, there is every potential for the establishment of a compost tea's own supply chain of fungicides which ultimately would lead to global acceptance on the application of compost teas in crop protection area. It is imperative to understand the microbial dynamics inside the compost teas as well as the antagonistic interaction with phytopathogens in plant. It is also noteworthy to look into other relevant strategies associated with sustainable farming practices such as crop rotation, minimal use of tillage, and organic amendments in view of safeguarding plant health in herbal cropping systems in relation to disease risk.

References

Abas F, Lajis NH, Israf DA, Khozirah S, Kalsom YU (2006) Antioxidant and nitric oxide inhibition activities of selected Malay traditional vegetables. Food Chem 95:566–573

Aghamohammadi Z, Etesami H, Alikhani HA (2016) Vermiwash allows reduced application rates of acaricide azocyclotin for control of two spotted spider mite, *Tetranychus urticae* Koch, on bean plant (*Phaseolus vulgaris* L.) Ecol Eng 93:234–241

Akila R, Rajendran L, Harish S, Saveetha K, Raguchander T, Samiyappan R (2011) Combined application of botanical formulations and biocontrol agents for the management of *Fusarium oxysporum* f. sp. *cubense* (Foc) causing Fusarium wilt in banana. Biol Control 57:175–183

Albert N, Nazaire K, Hartmut K (2012) The relative effects of compost and non-aerated compost tea in reducing disease symptoms and improving tuberization of *Solanum tuberosum* in the field. Int J Agric Res Rev 2:504–512

Allen P (1994) Accumulation profiles of lead and the influence of cadmium and mercury in *Oreochromis aureus* (Steindachner) during chronic exposure. Toxicol Environ Chem 44:101–112

Al-Mughrabi KI, Bertheleme C, Livingston T, Burgoyne A, Poirier R, Vikram A (2008) Aerobic compost tea, compost, and a combination of both reduce the severity of common scab (*Streptomyces scabiei*) on potato tubers. J Plant Sci 3:168–175

Bernal-Vicente A, Ros M, Tittarelli, Intrigliolo F, Pascual JA (2008) Citrus compost and its water extract for cultivation of melon plants in greenhouse nurseries. Evaluation of nutriactive and biocontrol effects. Bioresour Technol 99:8722–8728

Borrero C, Trillas M, Delgado A, Aviles M (2012) Effect of ammonium/nitrate ratio in nutrient solution on control of fusarium wilt of tomato by *Trichoderma asperellum* T34. Plant Pathol 61:132–139

Brinton WF, Droffner M (1995) The control of plant pathogenic fungi by use of compost teas. Biodynamics 197:12–15

Carballo T, Gil MV, Gomez X, Gonzalez-Andres F, Moran A (2008) Characterization of different compost extracts using Fourier-transform infrared spectroscopy (FTIR) and thermal analysis. Biodegradation 19:815–830

Castano R, Borrero C, Aviles M (2011) Organic matter fractions by SP-MAS ^{13}C NMR and microbial communities involved in the suppression of Fusarium wilt in organic growth media. Biol Control 58:286–293

Chae DH, Jin RD, Hwanagbo H, Kim YW, Kim YC, Park RD, Krishnan HB, Kim KY (2006) Control of late blight (*Phytophthora capsici*) in pepper plant with compost containing multitude of chitinase producing bacteria. BioControl 51:339–351

Dianez F, Santos M, Boix A, de Cara M, Trillas I, Aviles M, Tello JC (2006) Grape marc compost tea suppressiveness to plant pathogenic fungi: role of siderophores. Compost Sci Util 14:48–53

Dionne A, Tweddell RJ, Antoun H, Avis TJ (2012) Effect of non-aerated compost teas on damping-off pathogens of tomato. Can J Plant Pathol 34:51–57

Duffy B, Sarreal C, Ravva S, Stanker L (2004) Effect of molasses on regrowth of *E. coli* O157:H7 and *Salmonella* in compost teas. Compost Sci Util 12:93–96

Dukare AS, Prasanna R, Dubey SC, Nain L, Chaudhary V, Singh R, Saxena AK (2011) Evaluating novel microbe amended composts as biocontrol agents in tomato. Crop Prot 30:436–442

Elad Y, Freeman S (2002) Biological control of fungal plant pathogens. In: Kempken F (ed) The Mycota A comprehensive treatise on fungi as experimental systems for basic and applied research. Agricultural applications. Springer, Heidelberg, pp 93–109

El-Din AAE, Hendawy SF (2010) Effect of dry yeast and compost tea on growth and oil content of *Borago officinalis* plant. Res J Agric Biol Sci 6:424–430

El-Masry MH, Khalil AI, Hassouna MS, Ibrahim HAH (2002) In situ and in vitro suppressive effect of agricultural composts and their extracts on some phytopathogenic fungi. World J Microbiol Biotechnol 18:551–558

Esfahani MN, Monazzah M (2011) Identification and assessment of fungal diseases of major medicinal plants. J Ornamental Horticult Plants 1:137–145

Evans KJ, Palmer AK, Metcalf DA (2013) Effect of aerated compost tea on grapevine powdery mildew, botrytis bunch rot and microbial abundance on leaves. Eur J Plant Pathol 135:661–673

Eyheraguibel B, Silvestre J, Morard P (2008) Effects of humic substances derived from organic waste enhancement on the growth and mineral nutrition of maize. Bioresour Technol 99:4206–4212

Fritz JI, Whittle-Franke IH, Haindl S, Insam H, Braun R (2011) Microbiological community analysis of vermicompost tea and its influence on the growth of vegetables and cereals. Can J Microbiol 58:836–847

Gava CAT, Pinto JM (2016) Biocontrol of melon wilt caused by *Fusarium oxysporum* Schlecht f. sp. *melonis* using seed treatment with *Trichoderma* spp. and liquid compost. Biol Control 97:13–20

Gimsing AL, Kirkegaard JA (2009) Glucosinolates and biofumigation: fate of glucosinolates and their hydrolysis products in soil. Phytochem Rev 8:299–310

Haggag WM, Saber MSM (2007) Suppression of early blight on tomato and purple blight on onion by foliar sprays of aerated and non-aerated compost teas. J Food Agric Environ 5:302–309

Haran S, Schickler A, Oppenheim A, Chet I (1996) Differential expression of *Trichoderma harzianum* chitinase during mycoparasitism. Phytopathology 86:980–985

Hariprasad P, Divakara ST, Niranjana SR (2011) Isolation and characterization of chitinolytic rhizobacteria for the management of Fusarium wilt in tomato. Crop Prot 30:1606–1612

Hassan MAE, Abo-Elyousr KAM (2013) Impact of compost application on Fusarium wilt disease incidence and microelements contents of basil plants. Arch Phytopathol Plant Protect 46:1904–1918

Hegazy MI, Hussein EI, Ali AS (2015) Improving physiochemical and microbiological quality of compost tea using different treatments during extraction. Afr J Microbiol Res 9:763–770

Ingham E, Alms M (1999) Compost tea handbook. Soil Foodweb, Corvallis

Ingram D, Millner P (2007) Factors affecting compost tea as a potential source of *Escherichia coli* and *Salmonella* on fresh produce. J Food Prot 70:828–834

Islam M, Mondal C, Hossain I, Meah M (2014) Compost tea as soil drench: an alternative approach to control bacterial wilt in brinjal. Arch Phytopathol Plant Protect 47:1475–1488

Islam MK, Yaseen T, Traversa A, Ben Kheder M, Brunnetti G, Cocozza C (2016) Effects of the main extraction parameters on chemical and microbial characteristics of compost tea. Waste Manag 52:62–68

Janisiewicz WJ, Korsten L (2002) Biological control of postharvest diseases of fruits. Annu Rev Phytopathol 40:411–441

Javanmardi J, Ghorbani E (2012) Effects of chicken manure and vermicompost teas on herb yield, secondary metabolites and antioxidant activity of lemon basil (*Ocimum* x *citriodorum* Vis.) Adv Hortic Sci 26:151–157

Jinantana J, Sariah M (1998) Potential for biological control of *Sclerotium* foot rot of chili by *Trichoderma* spp. Pertanika J Trop Agric Sci 21:1–10

Joshi D, Hooda KS, Bhatt JC, Mina BL, Gupta HS (2009) Suppressive effects of composts on soil-borne and foliar diseases of French bean in the field in the western Indian Himalayas. Crop Prot 28:608–615

Kabir SME, Islam MR, Khan MMR, Hossain I (2012) Comparative efficacy of compost, poultry litter, IPM lab biopesticide and BAU biofungicide in controlling early blight of tomato. Int Res J Appl Life Sci 1:27–36

Kalra A, Singh HB, Patra NK, Kumar S (2003) Integrated host plant resistance and fungicide application on leaf blight control in menthol mint (Mentha arvensis L.) J Herbs Spices Med Plants 10:83–87

Kannangara T, Forge T, Dang B (2006) Effects of aeration, molasses, kelp, compost type and carrot juice on the growth of *Escherichia coli* in compost teas. Compost Sci Util 14:40–47

Katan J (2015) Soil solarization: the idea, the research and its development. Phytoparasitica 43:1–4

Kone SB, Dionne A, Tweddell RJ, Antoun H, Avis TJ (2010) Suppressive effect of non-aerated compost teas on foliar fungal pathogens of tomato. Biol Control 52:167–173

Lagerlof J, Insunza V, Lundegardh B, Ramert B (2011) Interaction between a fungal plant disease, fungivorous nematodes and compost suppressiveness. Acta Agric Scand Sect B-Soil Plant Sci 61:372–377

Litterick A, Wood M (2009) The use of composts and compost extracts in plant disease control. In: Walters D (ed) Disease control in crops: biological and environmentally friendly approaches. Wiley-Blackwell, Oxford, pp 93–121

Litterick AM, Harrier L, Wallace P, Watson CA, Wood M (2004) The role of uncomposted materials, composts, manures, and compost extracts in reducing pest and disease incidence and severity in sustainable temperate agricultural and horticultural crop production-a review. Crit Rev Plant Sci 23:453–479

Manandhar T, Yami KD (2008) Biological control of foot rot disease of rice using fermented products of compost and vermicompost. Sci World 6:52–57

Marin F, Santos M, Dianez F, Carretero F, Gea FJ, Yau JA, Navarro MJ (2013) Characters of compost teas from different sources and their suppressive effect on fungal phytopathogens. World J Microbiol Biotechnol 29:1371–1381

McQuilken MP, Whipps JM, Lynch JM (1994) Effects of water extracts of a composted manure-straw mixture on the plant pathogen *Botrytis cinerea*. World J Microbiol Biotechnol 10:20–26

Mehta CM, Palni U, Franke-Whittle IH, Sharma AK (2014) Compost: its role, mechanism and impact on reducing soil-borne plant diseases. Waste Manag 34:607–622

Merrill R, Mckeon J (2001) Apparatus design and experimental protocol for organic compost teas. Org Farming Res Found 9:9–15

Messiha NAS, van Diepeningen AD, Wenneker M, van Beuningen AR, Janse JD, Coenen TGC, Termorshuizen AJ, van Bruggen AHC, Blok WJ (2007) Biological soil disinfestation (BDS), a new control method for potato brown rot caused by *Ralstonia solanacearum* race 3 biovar 2. Eur J Plant Pathol 117:403–415

Mitra J, Bhuvaneshwari V, Paul PK (2013) Broad spectrum management of plant diseases by phylloplane microfungal metabolites. Arch Phytopathol Plant Protect 46:1993–2001

Mohd Din ARJ, Hanapi SZ, Sarmidi MR (2016) Suppressive effect of compost tea on fungal pathogen causing melanose on *Melicope ptelefolia* (Tenggek Burung) leaf. 6th International conference of biotechnology for wellness industries. Melaka, pp 5–8

Moosa A, Sahi ST, Haq I, Farzand A, Khan SA, Javaid K (2016) Antagonistic potential of Trichoderma isolates and manures against Fusarium wilt of tomato. Int J Veg Sci 1–12

Muscolo A, Sidari M (2009) Carboxyl and phenolic humic fractions affect callus growth and metabolism. Soil Sci Soc Am J 73:1119–1129

Naidu Y, Sariah M, Jugah K, Siddiqui Y (2010) Microbial starter for the enhancement of biological activity of compost tea. Int J Agric Biol 12:51–56

Naidu Y, Meon S, Siddiqui Y (2013) Foliar application of microbial-enriched compost tea enhances growth, yield and quality of muskmelon (*Cucumis melo* L.) cultivated under fertigation system. Sci Hortic 159:33–40

Nobel R, Coventry E (2005) Suppression of soil-borne plant disease with composts: a review. Biocontrol Sci Tech 15:3–20

Olawuyi OJ, Odebode AC, Oyewole IO, Akanmu AO, Afolabi O (2014) Effect of arbuscular mycorrhizal fungi on *Pythium aphanidermatum* causing foot rot disease on pawpaw (*Carica papaya* L.) seedlings. Arch Phytopathol Plant Protect 47:185–193

On A, Wong F, Ko Q, Tweddell RJ, Antoun H, Avis TJ (2015) Antifungal effects of compost tea microorganisms on tomato pathogens. Biol Control 80:63–69

Palmer AK, Evans KJ, Metcalf DA (2010) Characters of aerated compost tea from immature compost that limit colonization of bean leaflets by *Botrytis cinerea*. J Appl Microbiol 109:1619–1631

Pane C, Piccolo A, Spaccini R, Celano G, Villecco D, Zaccardelli M (2013) Agricultural waste-based composts exhibiting suppressivity to diseases caused by the phytopathogenic soil-borne fungi *Rhizoctonia solani* and *Sclerotinia minor*. Appl Soil Ecol 65:43–51

Pane C, Celano G, Zaccardelli M (2014) Metabolic patterns of bacterial communities in aerobic compost teas associated with potential biocontrol of soilborne plant diseases. Phytopathol Mediterr 53:277–286

Pane C, Palese AM, Spaccini R, Piccolo A, Celano G, Zaccardelli M (2016) Enhancing sustainability of a processing tomato cultivation system by using bioactive compost teas. Sci Hortic 202:117–124

Pant AP, Radovich TJK, Hue NV, Paull RE (2012) Biochemical properties of compost tea associated with compost quality & effects on pak choi growth. Sci Hortic 48:138–146

Raj D, Antil RS (2011) Evaluation of maturity and stability parameters of compost prepared from agro-industrial wastes. Bioresour Technol 102:2868–2873

Romero D, Perez-Garcia A, Rivera ME, Cazorla FM, de Vicente A (2004) Isolation and evaluation of antagonistic bacteria towards the cucurbit powdery mildew fungus *Podosphaera fusca*. Appl Microbiol Biotechnol 64:263–269

Saha M, Sarkar S, Sarkar B, Sharma BK, Bhattacharjee S, Tribedi P (2016) Microbial siderophores and their potential applications: a review. Environ Sci Pollut Res 23:3984–3999

Said-Pullicino D, Erriquens FG, Gigliotti G (2007) Changes in the chemical characteristics of water-extractable OM during composting and their influence on compost stability and maturity. Bioresour Technol 98:1822–1831

Sang MK, Kim KD (2011) Biocontrol activity and primed systemic resistance by compost water extracts against anthracnoses of pepper and cucumber. Phytopathology 101:732–740

Sang MK, Kim J, Kim KD (2010) Biocontrol activity and induction of systemic resistance in pepper by compost water extracts against *Phytophthora capsici*. Biol Control 100:774–783

Santos M, Dianez F, del Valle MG, Tello JC (2008) Grape marc compost: microbial studies and suppression of soil-borne mycosis in vegetable seedlings. World J Microbiol Biotechnol 24:1493–1505

Scheuerell SJ, Mahaffee WF (2002) Compost tea: principles and prospects for plant disease control. Compost Sci Util 10:313–338

Scheuerell SJ, Mahaffee WF (2004) Compost tea as a container medium drench for suppressing seedling damping-off caused by *Pythium ultimum*. Phytopathology 94:1156–1163

Scheuerell SJ, Mahaffee WF (2006) Variability associated with suppression of gray mold (*Botrytis cinerea*) on geranium by foliar applications of non-aerated and aerated compost teas. Plant Dis 90:1201–1208

Seddigh S, Kiani L, Tafaghodinia B, Hashemi B (2014) Using aerated compost tea in comparison with a chemical pesticide for controlling rose powdery mildew. Arch Phytopathol Plant Protect 47:658–664

Segarra G, Reis M, Casanova E, Trillas I (2009) Control of powdery mildew (*Erysiphe polygoni*) in tomato by foliar applications of compost tea. J Plant Pathol 91:683–689

Senesi N, Plaza C, Brunetti G, Polo A (2007) A comparative survey of recent results on humic-like fractions in organic amendments and effects on native soil humic substances. Soil Biol Biochem 39:1244–1262

Shrestha K, Adetutu EM, Shrestha P, Walsh KB, Harrower KM, Ball AS, Midmore DJ (2011a) Comparison of microbially enhanced compost extracts produced from composted cattle rumen content material and from commercially available inocula. Bioresour Technol 102:7994–8002

Shrestha K, Shrestha P, Walsh KB, Harrower KM, Midmore DJ (2011b) Microbial enhancement of compost extracts based on cattle rumen content compost-characterisation of a system. Bioresour Technol 102:8027–8034

Siddiqui Y, Sariah M, Ismail MR, Asgar A (2008) *Trichoderma*-fortified compost extracts for the control of Choanephora wet rot in okra production. Crop Prot 27:385–390

Siddiqui Y, Meon S, Ismail R, Rahmani M (2009) Bio-potential of compost tea from agro-waste to suppress *Choanephora cucurbitarum* L. the causal pathogen of wet rot of okra. Biol Control 49:38–44

Siddiqui Y, Islam TM, Naidu Y, Meon S (2011) The conjunctive use of compost tea and inorganic fertilizer on the growth, yield and terpenoid content of *Centella asiatica* (L.) urban. Sci Hortic 130:289–295

St. Martin CCG, Dorinvil W, Brathwaite RAI, Ramsubhag A (2012) Effects and relationships of compost type, aeration and brewing time on compost tea properties, efficacy against *Pythium ultimum*, phytotoxicity and potential as a nutrient amendment for seedling production. Biol Agric Hortic:1–21

Trankner A (1992) Use of agricultural and municipal organic wastes to develop suppressiveness to plant pathogens. In: Tjamos ES, Papavizas GC, Cook RJ (eds) Biological control of plant diseases. Plenum Press, New York, pp 35–42

Wahyuni WS, Mudjiharjati A, Sulistyaningsih N (2010) Compost extracts of vegetable wastes as biopesticide to control cucumber mosaic virus. J Biosci 17:95–100

Weltzien HC (1991) Biocontrol of foliar fungal diseases with compost extracts. In: Andrews JH, Hirano SS (eds) Microbial ecology of leaves. Springer, New York, pp 430–450

Xu D, Raza W, Yu G, Zhao Q, Shen Q, Huang Q (2012) Phytotoxicity analysis of extracts from compost and their ability to inhibit soil-borne pathogenic fungi and reduce root-knot nematodes. World J Microbiol Biotechnol 28:1193–1201

Xu D, Zhao S, Xiong Y, Peng C, Xu X, Si G, Yuan J, Huang Q (2015) Biological, physiochemical and spectral properties of aerated compost extracts: influence of aeration quantity. Commun Soil Sci Plant Anal 46:2295–2310

Yohalem D, Harris R, Andrews J (1994) Aqueous extracts of spent mushroom substrate for foliar disease control. Compost Sci Util 2:67–74

Yohalem DS, Voland R, Nordheim EV, Harris RF, Andrews JH (1996) Sample size requirements to evaluate spore germination inhibition by compost extracts. Soil Biol Biochem 28:519–525

Zhang W, Han DY, Dick WA, Davis KR, Hoitink HAJ (1998) Compost and compost water extract-induced systemic acquired resistance in cucumber and Arabidopsis. Phytopathology 88:45–545

Zhou Y, Selvam A, Wong JWC (2016) Effect of Chinese medicinal herbal residues on microbial community succession and anti-pathogenic properties during co-composting with food waste. Bioresour Technol 217:190–199

Zmora-Nahum S, Danon M, Hadar Y, Chen Y (2008) Chemical properties of compost extracts inhibitory to germination of *Sclerotium rolfsii*. Soil Biol Biochem 40:2523–2529

Chapter 9
Enzymatic Hydrolysis of Used Cooking Oil Using Immobilized Lipase

Nor Athirah Zaharudin, Roslina Rashid, Lianash Azman, Siti Marsilawati Mohamed Esivan, Ani Idris, and Norasikin Othman

Abstract The increasing production of used cooking oil (UO) has contributed to serious environmental problems. Management of this oil posed a significant challenge especially on choosing the right disposal method considering the possibilities of water and land resources contamination. Abundant amount of UO from food industry and household draw an attention where this waste could be utilized and recycled to produce valuable products. Currently, UO has been utilized for the production of biodiesel. Enzymatic hydrolysis of UO using lipase is one of the promising approaches to produce free fatty acids and glycerol, both of which are highly utilized especially in the oleochemical industries. In addition, enzymatic hydrolysis offers lots of advantages compared to conventional fat splitting process, notably its lower reaction temperature and high substrate specificity leading to products with high purity and low by-products.

N.A. Zaharudin • R. Rashid (✉) • L. Azman • S.M.M. Esivan
Department of Bioprocess and Polymer Engineering, Faculty of Chemical and Energy Engineering, Universiti Teknologi Malaysia, Johor Bahru, Johor, Malaysia
e-mail: r-roslina@utm.my

A. Idris
Department of Bioprocess and Polymer Engineering, Faculty of Chemical and Energy Engineering, Universiti Teknologi Malaysia, Johor Bahru, Johor, Malaysia

Institute of Bioproduct Development, Universiti Teknologi Malaysia, Johor Bahru, Johor, Malaysia
e-mail: ani@cheme.utm.my

N. Othman
Department of Chemical Engineering, Faculty of Chemical and Energy Engineering, Universiti Teknologi Malaysia, Johor Bahru, Johor, Malaysia

Z.A. Zakaria (ed.), *Sustainable Technologies for the Management of Agricultural Wastes*, Applied Environmental Science and Engineering for a Sustainable Future, https://doi.org/10.1007/978-981-10-5062-6_9

9.1 Introduction

The increase in human population has caused rapid development in economic sectors that indirectly contributes to the increase in used cooking oil (UO) generation. More restaurants and fast-food centres have expended and offered various types of foods that use vegetable oils during the preparation process. Massive amount of UOs was unreasonably discharged to the environment causing environmental pollutions, while potential for long-term consumption of recycled UO may lead to human health problems. Existing studies on UO are mainly focusing on its utilization for the production of biodiesel and soap. However, utilization of UOs is still very limited and lacking in scientific investigation. With proper pretreatment and appropriate process, UO offers an opportunity to be a good raw material for various industries including oleochemicals and machinery.

Generally, enzymatic hydrolysis of UO produced free fatty acids and glycerol. Fatty acids and glycerol are essential raw materials for food, cosmetics, and pharmaceutical industries. Typical production of fatty acids and glycerol is from the hydrolysis of fresh cooking oils. High content of free fatty acids in UO compared to fresh cooking oil indicates that high free fatty acids recovery could be achieved with suitable process. Similarity in quality of free fatty acids produced from UO and fresh cooking oil could be accomplished at optimum reaction conditions. Conventional production of fatty acids from fresh cooking oils proceeds at high temperature and high pressure. Enzymatic hydrolysis requires mild reaction temperature and pressure with less or no side products, which generate a lesser expenditure on energy consumption and downstream processes.

9.2 The Potential Use of Used Cooking Oil for Fatty Acids Production

9.2.1 Introduction

Developed countries are the major contributors to UO. Maddikeri et al. (2012) reported that about 29 million tons of UO are generated every year. Kalam et al. (2011) reported that 0.5 million tons of UO is produced in Malaysia every year. With the mushrooming establishment of fast-food centres and restaurants, amount of UOs are expected to increase tremendously. Sebayang et al. (2010) estimated the total UO generated from fast-food franchise all over Malaysia reach several thousand litres a day. It is believed that the actual amount of UO generated is far beyond the estimated value. Nurdin et al. (2016) reported that Malaysia has produced approximately 5000 tons of UO derived from vegetable oils and animal fats are disposed without proper treatment. The availability of UO in Malaysia could provide a stable and constant feedstock to meet the demand on fatty acids and glycerol production.

Using UO as raw material for fatty acid production not only reduces the cost but also addresses a crucial environmental problem as well as provides a solid foundation for the utilization of UO. Compared to fresh cooking oils, the cost of UOs is much lower ranging anywhere no cost to less than 60% (relative to fresh cooking oils), depending on the source and availability (Predojevic 2008). Phan and Phan (2008) indicate that the price of UO is two to three times cheaper than fresh vegetable oil. Waghmare and Rathod (2016) stated that utilization of UOs is for the production of surfactant, bio-lubricants, soap, synthetic detergents, greases, cosmetics, and biodiesel. Even though some of UO is being utilized in many sectors, still major portion of it is constantly discharged into the environment.

Enweremadu and Mbarawa (2009) stated that there is no specific or systematic processing method for UOs. Most of the oils is thrown through home drains, ends up in wastewaters, and then is discharged as surface waters, resulting in water pollutions. More than 80% of the oil is consumed at home. Due to large volume involved, it is extremely difficult to control the disposal behaviour of public (Alcantara et al. 2000). The insight of UO exploitation for biodiesel production is generally from the argument of energy-crop programmes versus food crops that in long term will cause serious food shortage and increase in price especially in developing country. Hence, the utilization of UO for free fatty acid production will bring significant environmental profits as well as offers a new solution for final disposal of the oil which contributes to the alleviation of environmental problems.

9.2.2 Chemical and Physical Properties of UO

UOs are commonly generated from vegetable oils that were used repeatedly at high temperature, especially for frying or food preparation purposes. In general, vegetable oils will undergo various chemical and physical changes during the frying process (Kulkarni and Dalai 2006). During the frying process, oil is heated with the occurrence of air and light at temperature of 160–200 °C for extended time (Gui et al. 2008; Kulkarni and Dalai 2006). Five common physical changes that occurred in cooking oil during the frying process are increase in viscosity, change in surface tension, increase in specific heat, colour changing, and also the increase in fat foam content. In addition, common degradation reactions that occurred are hydrolysis, oxidation, and polymerization (Sanli et al. 2011). Combination of all the reactions initiates the formation of undesirable products as well as increases the content of polar materials in the oil.

Continuous frying process resulted in the increase in saturation degree and change in fatty acid composition. Oxidation, polymerization, and scission reactions are the main contributors towards the rupture of the double bond as the oil is heated (Alireza et al. 2010). In addition, the changes also influence some of the oil properties such as iodine value, viscosity, and density depending on frequency of the frying process (Zaharudin et al. 2016). According to Tynek et al. (2001), a significant loss of linoleic acid (C18:2) and decrease in iodine value of the oil are caused by

Table 9.1 Chemical properties of UPO (Abdul Halim and Kamaruddin 2008; Alireza et al. 2010; Tomasevica and Siler-Marinkovic 2003)

Compound	Fresh cooking oil	Used frying oil
Water content (% wt)	0.8	0.95
Acid (mg KOH/g)	0.1	0.1–17.85
Peroxide value (meq/kg)	3.4	13.2–66
Iodine value (g I_2/g oil)	57.27	44.6–52.81
FFA (%)	0.13	0.2

Table 9.2 Fatty acid composition of fresh cooking palm oil and UPO (Taufiqurrahmi et al. 2011; Dauqan et al. 2011)

Fatty acids	Carbon number: double bond	Fresh cooking palm oil (wt%)	Used cooking palm oil (wt%)
Oleic acid	C18:1	49.482	28.64
Palmitic acid	C16:0	36.768	21.47
Linoleic acid	C18:2	11.747	13.58
Stearic acid	C18:0	–	13.00
Palmitoleic acid	C16:1	–	7.56
Myristic acid	C14:0	0.849	3.21
Linolenic acid	C18:3	0.539	1.59
Lauric acid	C12:0	0.230	1.1
Arachidic acid	C20:0	0.161	0.64
Heptadecanoic acid	C17:0	–	0.59
Arachidonic acid	C20:4	–	0.37
Eicosadienoic acid	C20:2	–	0.29
Others		0.219	8.04

intense thermo-oxidative transformation. Choe and Min (2007) indicate that heating oil will cause oxidative rancidity which leads to an increase in free fatty acid content. Study conducted by Taufiqurrahmi et al. (2011) showed that used cooking palm oil consists of approximately 28% of oleic acid and 21.47% of palmitic acid, which are much lower values compared to fresh cooking oil, with values of 49.48% (oleic acid) and 36.77% (palmitic acid), respectively. Table 9.1 shows chemical properties in fresh cooking oil and UO, while Table 9.2 shows fatty acid compositions of fresh cooking palm oil and used cooking palm oil.

9.3 Enzyme: Lipase

9.3.1 *Introduction*

Generally, lipase is known as triacylglycerol (TAG) acylhydrolyses (EC 3.1.1.3). It is a form of carboxy esterase which catalyses hydrolysis of oils and fats and reacts without the presence of cofactor. Lipases are enzymes that break down lipids (the general name given for fats and oils) into glycerol and fatty acids. The natural substrates for lipase are triglycerols, which have low solubility in water. Lipases have been used extensively as biocatalyst for the alteration of fats and oils with the presence of artificial chemicals under controlled conditions in aqueous and nonaqueous media. Lipases have been widely used in various industrial applications due to the ability of utilizing various range of substrates, high stability in extreme temperature, pH, and organic solvent as well as chemo-, region- and enantio-selectivities (Saxena et al. 2003). In addition, lipases were isolated from diverse sources of bacterial, fungal, and animal which have inclusive range of positional specificity, fatty acids specificity, thermostability, optimum pH, and others (Joseph et al. 2008).

9.3.2 *Lipase Properties*

Lipases have different pH ranges depending on its origin of isolation. For example, pH range for plant lipase is 4.0–9.0; animal lipase, 5.5–8.5; and microorganisms, 6.0–10.0. Mobarak-Qamsari et al. (2011) described that bacterial lipases have a neutral or alkaline optimum pH, while fungal lipases have broad range of optimal pH ranging from 5.5 to 10.5. As for temperature, plant lipase remains active between 20 and 38 °C, while for animal lipase, the optimum temperature was reported to be between 37 and 60 °C. The optimum temperature for microbial lipase varies between 37 and 55 °C (Pahoja and Sethar 2002). The plant and animal lipases are active in the presence of Ca and Zn, while the activity would be inhibited in the presence of EDTA, Triton X-100, and Tween 80. However, microbial-based lipases are inhibited in the presence of $FeCl_3$, $ZnCl_2$, and $HgCl_2$ (Pahoja and Sethar 2002).

9.3.2.1 Immobilization of Lipase

Commercially available lipases are available in both free and immobilized forms. As for this current study, immobilized lipase was used for the hydrolysis of used frying oil. Da Silva et al. (2008) stated that immobilized lipases have higher activity compared to free-form lipases and demonstrate stabilization of enzymatic protein conformation as well as enhanced thermal denaturation. Sayin et al. (2011), on the other hand, also proved that when *C. rugosa* lipase was immobilized on alkyl

N-methylglucamine, it shows high activity, is easily separated and stored, as well as has the ability to retain activity after several usages.

Ting et al. (2006) reported that the immobilization of *C. rugosa* lipase on chitosan beads resulted in the increase of both thermal and pH stability of the enzymes. The optimum temperature of lipase has shifted from 35 to 40 °C. As for pH, the optimum pH shifted from 7.0 to 8.0 once the immobilization process was completed. Liu et al. (2005) observed a shift in optimum pH (from 7 to 8) and temperature (from 37 to 50 °C) during the hydrolysis of olive oil using lipase immobilized onto micron-sized magnetic beads. On the other hand, Chang et al. (2007) successfully immobilized *C. rugosa* lipase on Celite-545 in order to obtain the optimum hydrolysis conditions using response surface methodology (RSM) and factorial experimental design. The study shows that immobilization of lipases offer diverse organic syntheses as it permits simple immobilization of lipase through inexpensive process. Murty et al. (2002) described that the use of immobilized lipase could potentially reduce up to 20% of the cost (relative to the use of traditional free-form enzyme) where most of the savings were attributed to the cost of acquiring nonreusable enzymes. In addition, immobilization also improves the overall process control as well as product quality.

9.3.3 Catalytic Mechanism

Generally, lipase catalysed three types of reactions, including hydrolysis of triglycerides, ester synthesis, and trans-esterification. On the other hand, trans-esterification can be further classified into four classes related to the chemical compound that react with ester, which includes alcoholysis, inter-esterification, acidolysis, and aminolysis. The catalytic reaction of lipase is reversible. Lipase catalysed hydrolysis in aqueous system. However for the esterification (reverse reaction), lipase reacts in microaqueous system (low water content). The reverse reaction leads to the yield of glycerides from fatty acids and glycerol.

Hydrolysis using lipase is a process of acyl migration between glyceride molecules with the addition of water. Hydrolysis of oils or fats produces free fatty acids and glycerol. Salimon et al. (2011) conducted the optimization process of preparing fatty acids from *Jatropha curcas* seed oil. Gupta et al. (2011) have studied the hydrolysis reaction of different oils using membranes with embedded lipase in order to yield FFA. Figure 9.1 shows the catalytic reaction of lipase towards triglycerides to produce fatty acids and glycerol. Fatty acids and glycerol are products from the hydrolysis of TAG with lipase. Normally, fatty acids and glycerol are dissolved in the lipid-water phase. The reaction is reversible where the hydrolysis rate and the final composition significantly depend on the fatty acid concentration in the oil phase, with glycerol concentrate in the water phase (Murty et al. 2002).

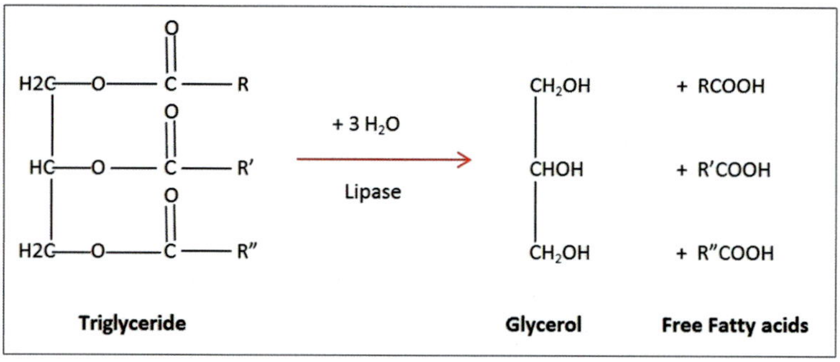

Fig. 9.1 Enzymatic hydrolysis of triglyceride

9.4 Case Study: Hydrolysis of Used Cooking Oil (UO) Using Immobilized Lipase

The following materials were used; UO was obtained from fast-food restaurants located around Taman Universiti, Johor, Malaysia, and immobilized lipase used (Sigma Aldrich) consists of lipase B *Candida antarctica* immobilized on Immobead 150 (recombinant from *Aspergillus oryzae*), aluminium sulphate $Al_2(SO_4)_3$, sodium carbonate (Na_2CO_3) 2% Tween-80 emulsifier, NaOH, phenolphthalein indicator, and phosphate buffer solution (mixture of monosodium phosphate, disodium phosphate, and phosphoric acid).

UO was pretreated using either $Al_2(SO_4)_3$ or Na_2CO_3. The treated oil was then analysed using gas chromatography to determine sample with higher composition of fatty acids. UO was heated at 40 °C prior to the addition of flocculants and $Al_2(SO_4)_3$ to remove impurities. The product from the reaction was allowed to stand to obtain the oil and water fractions. Small volume (10 mL) of the resulting oil fraction was mixed with 10 ml of phosphate buffer solution (pH 7.5) prior to transferring into a series of 250 ml conical flasks. The reaction was initiated by varying the enzyme load ranging from 0.2 to 1.0% (w/w). For each UO sample, three drops of 2% Tween-80 was added onto the mixture followed by mixing at 200 rpm, 45 °C for 75 mins. The extent of lipid hydrolysis by lipase B was evaluated by extracting 1 mL of reaction mixture every 15 mins. The effect of mixing speed on the hydrolysis process was monitored by varying the agitation speed between 150 and 300 rpm. To determine the effects of different parameters on hydrolysis degree, the default conditions were chosen as follows: enzyme loading of 0.8% (w/w), temperature of 45 °C, agitation speed of 300 rpm, and reaction time of 75 mins.

Amount of free fatty acids produced from hydrolysis reaction was quantified based on the hydrolysis degree. The hydrolysis degree was calculated using the following formula as stated by Rooney and Weatherly (2001).

$$P = \%\text{hydrolysis} = \frac{T \times 0.05 \times 10^{-3} \times \text{AMWFA}}{Wt \times fo} \times 100\%$$

where *T* is the volume of titration (ml), *AMWFA* is the average molecular weight of fatty acids (292.83 g/mol), *Wt* is the weight of sample (g), and *fo* is the fraction of oil at the start of reaction.

Figure 9.2 shows the effect of enzyme loading on the degree of hydrolysis. Degree of hydrolysis increased with the increase in enzyme loading, a situation similarly reported by Rashid et al. (2014). The highest percentage of hydrolysis degree, 63.43%, was achieved at enzyme loading of 0.8% and 1.0% (w/w). Enzyme loading of 0.8% (w/w) was selected as the optimum enzyme loading in this study compared to 1.0% (w/w) due to economic aspect as higher concentration of enzyme used in the reaction will contribute towards the increase in overall process cost. As the hydrolysis process operates without the presence of organic solvent, higher substrate concentration with greater production volume is required (Meng et al. 2011). Evaluation on optimum enzyme loading is crucial to avoid enzyme wastage. Goswami et al. (2009) indicated that in order to increase the degree of hydrolysis, it was necessary to continuously add high amount of lipase. However, high amount of lipase concentration would make the oil, water, and lipase mixture to form paste-like structure which would lead to difficulty in separation processes (Wang et al. 2011). Rooney and Weatherly (2001) obtained highest hydrolysis degree of sunflower oil at enzyme loading of 0.8% (w/w). Enzyme loading has a strong impact on the catalytic process. As a lipase amount increases, lipase moves from the aqueous phase to the interface at the increasing rate and the interaction with the substrate increases and results towards enhanced hydrolysis (Al-Zuhair et al. 2003).

The effect of agitation speed on the degree of hydrolysis is presented in Fig. 9.3. The degree of hydrolysis increased with the increase in agitation speed. The highest degree of hydrolysis was recorded at agitation speed of 300 rpm with 84.57% hydrolysis degree. Other literatures including Chew et al. (2008), Abdul Halim and Kamaruddin (2008), and Torres et al. (2007) also obtained highest hydrolysis degree

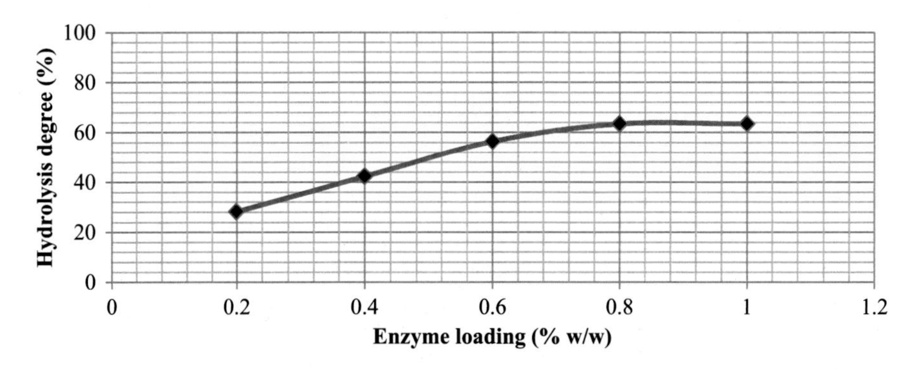

Fig. 9.2 Effect of enzyme loading on hydrolysis degree

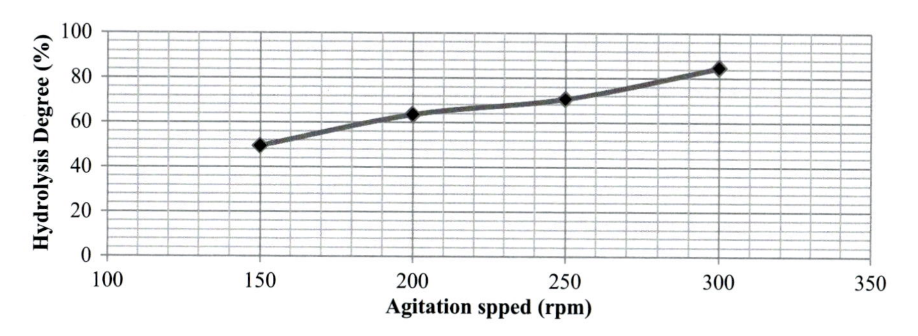

Fig. 9.3 Effect of agitation speed on hydrolysis degree

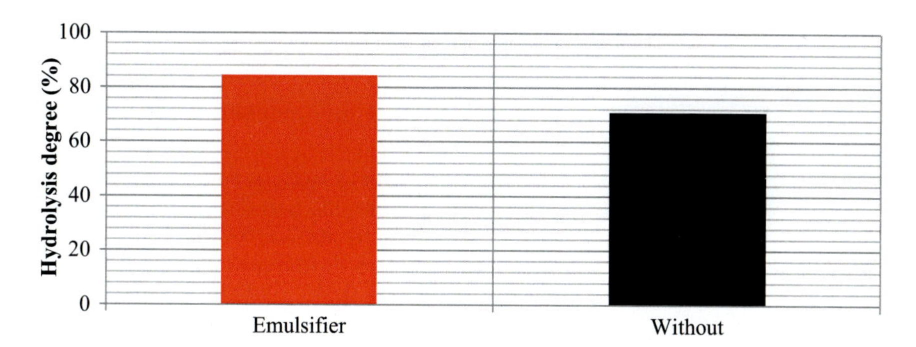

Fig. 9.4 Effect of emulsifier on hydrolysis degree

at agitation speed of 200 rpm for hydrolysis of different source of fresh edible oil. The optimum agitation speed established in the present study was higher compared to the reported value. Rashid et al. (2014) obtained higher optimum agitation speed of 220 rpm in the hydrolysis of UO. The differences in optimum agitation speeds are mainly due to the source of substrate and enzyme used. All the previous studies have been using fresh edible oil as a substrate. The optimum agitation speed for UO hydrolysis is 300 rpm with hydrolysis degree of 84.57%. Agitation speed influenced the hydrolysis degree as it reduces the droplet size, increasing the specific interfacial area between oil and the aqueous phase. Puthli et al. (2006) described that high agitation speed provides high shear force in the medium, generating a fine dispersion and providing large interfacial area for the hydrolysis reaction to occur.

Effect of emulsifier was determined at optimum enzyme loading of 0.8% (w/w), agitation speed of 300 rpm, temperature of 45 °C, and reaction time of 75 mins (Fig. 9.4). The positive role of emulsifier is clearly shown based on the higher degree of hydrolysis relative to without the presence of emulsifier. Emulsifier promotes the mixing of oil and water (Meng et al. 2011). The emulsifying properties of Tween-80 are based on their surface properties, as well as the effectiveness of the hydrolysate to lower the interfacial tension between the hydrophobic and hydrophilic components in oils. The mechanism of emulsification process is the absorption of proteins to the surface of newly formed oil droplets during homogenization and produces a protective membrane that avoids droplets from coalescing (Amiza et al. 2012). Hydrolysates are surface active materials that stimulate oil-in-water emulsion due to their water-soluble properties containing hydrophilic and hydrophobic functional groups (Groninger and Miller 1979).

As a conclusion, enzymatic hydrolysis of UO is clearly affected by the micro-environment introduced into the system. Optimum reaction parameters ensure high production of free fatty acids as well as avoiding wastage of substrates and enzyme usage which leads to the increase in overall cost. The production of free fatty acids from hydrolysis of UO using immobilized lipase has been successfully restored in solvent-free system. Enzymatic hydrolysis of UO for free fatty acid production is a promising method for environment-friendly treatment for utilization of UO.

References

Abdul Halim SF, Kamaruddin AH (2008) Catalytic studies of lipase on FAME production from waste cooking palm oil in a tert-butanol system. J Process Biochem 43:1436–1439

Alcantara R, Amores J, Canoira L, Fidalgo E, Franco MJ, Navarro A (2000) Catalytic production of biodiesel from soy-bean oil, used frying oil and tallow. Biomass Bioenergy 18:515–527

Alireza S, Tan CP, Hamed M, Che Man YB (2010) Effect of frying process on fatty acid composition and iodine value of selected vegetable oils and their blends. Int Food Res J 17:295–302

Al-Zuhair S, Hasan M, Ramachandran KB (2003) Kinetic hydrolysis of palm oil using lipase. J Process Biochem 38:1155–1163

Amiza MA, Kong YL, Faazaz AL (2012) Effects of degree of hydrolysis on physicochemical properties of cobia (Rachycentron canadum) frame hydrolysate. Int Food Res J 19:199–206

Chang SF, Chang SW, Yen YH, Shieh CJ (2007) Optimum immobilization of Candida rugose lipase on celite by RSM. Appl Clay Sci J 37:67–73

Chew YH, Chua LS, Cheng KK, Sarmidi MR, Aziz RA, Lee CT (2008) Kinetic study on the hydrolysis of palm olein using immobilized lipase. Biochem Eng J 39:516–520

Choe E, Min V (2007) Chemistry of deep-fat frying oils. J Food Sci 72:77–86

da Silva VCF, Contesini FJ, de O. Carvalho P (2008) Characterization and catalytic activity of free and immobilized lipase from Aspergillus niger: a comparative study. J Braz Chem Soc 19:1468–1474

Dauqan EMA, Abdullah Sani H, Abdullah A, Mohd Kasim Z (2011) Fatty acids composition of four different vegetable oils (red palm olein, palm olein, corn oil, and coconut oil) by gas chromatography, pp 29–31

Enweremadu CC, Mbarawa MM (2009) Technical aspects of production and analysis of biodiesel from used cooking oil—a review. Renew Sust Energ Rev 13:2205–2224

Goswami D, Basu JK, De S (2009) Optimization of process variables in castor oil hydrolysis by Candida rugosa lipase with buffer as dispersion medium. Biotechnol Bioprocess Eng 14:220–224

Groninger HS, Miller R (1979) Some chemical and nutritional properties of acylated fish proteins. J Agric Food Chem 27:948–955

Gui MM, Lee KT, Bhatia S (2008) Feasibility of edible oil vs. non-edible oil vs. waste edible oil as biodiesel feedstock. Energy 33:1646–1653

Gupta S, Ingole P, Singh K, Bhattacharya A (2011) Comparative study of the hydrolysis of different oils by lipase-immobilized membranes. J Appl Polym Sci 124:17–26
Joseph B, Ramteke PW, Thomas G (2008) Cold active microbial lipases: some hot issues and recent developments. Biotechnol Adv 26:457–470
Kalam MA, Masjuki HH, Jayed MH, Liaquat AM (2011) Emission and performance characteristics of an indirect ignition diesel engine fuelled with waste cooking oil. Energy 36:397–402
Kulkarni MG, Dalai AK (2006) Waste cooking oil-an economical source for biodiesel: a review. Indian J Chem Eng 45:2901–2913
Liu X, Guan Y, Shen R, Liu H (2005) Immobilization of lipase onto micron-size magnetic beads. J Chromatogr B 822:91–97
Maddikeri GL, Pandit AB, Gogate PR (2012) Intensification approach for biodiesel synthesis from waste cooking oil: a review. Ind Eng Chem Res 51:14610–14628
Meng Y, Wang G, Yang N, Zhou Z, Li Y, Liang X, Chen J, Li Y, Li J (2011) Two-step synthesis of fatty acid ethyl ester from soybean oil catalysed by Yarrowia lipolytica lipase. Biotechnol Biofuels 4:1–9
Mobarak-Qamsari E, Kasra-Kermanshahi R, Moosavi-nejad Z (2011) Isolation and identification of a novel, lipase-producing bacterium, Pseudomonas aeruginosa KM110. Iran J Microbiol 3:92–98
Murty VR, Bhat J, Muniswaran PKA (2002) Hydrolysis of oils by using immobilized lipase enzyme: a review. Biotechnol Bioprocess Eng 7:57–66
Nurdin S, Yunus RM, Nour AH, Gimbun J, Azman NAN, Sivaguru MV (2016) Restoration of waste cooking oil (WCO) using lkaline hydrolysis technique (ALHT) for future biodetergent. ARPN J Eng Appl Sci 11:6405–6410
Pahoja VM, Sethar MA (2002) A review of enzymatic properties of lipase in plants, animals and microorganisms. J Appl Sci 2:474–484
Phan AN, Phan TM (2008) Biodiesel production from waste cooking oils. Fuel 87:3490–3496
Predojevic ZJ (2008) The production of biodiesel from waste frying oils: a comparison of different purification steps. Fuel 87:3522–3528
Puthli MS, Rathod VK, Pandit AB (2006) Enzymatic hydrolysis of castor oil: process intensification studies. Biochem Eng J 31:31–41
Rashid R, Zaharudin NA, Idris A (2014) Enzymatic hydrolysis of used-frying oil using Candida rugosa lipase. Jurnal Teknologi (Sci Eng) 67:101–107
Rooney D, Weatherly LR (2001) The effect of reaction conditions upon lipase catalysed hydrolysis of high oleate sunflower oil in a stirred liquid-liquid reactor. Process Biochem 36:947–953
Salimon J, Abdullah BM, Salih N (2011) Hydrolysis optimization and characterization study of preparing fatty acids from Jatropha curcas seed oil. Chem Cent J 5:1–9
Sanli H, Canakci M, Alptekin E (2011) Characterization of waste frying oils obtained from different facilities. World Renew Energy Congr. 2011 2011:479–485
Saxena RK, Sheoran A, Giri B, Davidson S (2003) Purification strategies for microbial lipases. J Microbiol Methods 52:1–18
Sayin S, Yilmaz E, Yilmaz M (2011) Improvement of catalytic properties of Candida rugosa lipase by sol–gel encapsulation in the presence of magnetic calix arene nanoparticles. Organic and. Biomol Chem 9:4021–4024
Sebayang D, Agustian E, Praptijanto A (2010)Transesterification of biodiesel from waste cooking oil using ultrasonic technique. International conference on environment, 13–15 December ICENV, Penang, pp 1–9
Taufiqurrahmi N, Mohamed AR, Bhatia S (2011) Production of biofuel from waste cooking palm oil using nanocrystalline zeolite as catalyst: process optimization studies. Bioresour Technol 102(22):10686–10694
Ting WJ, Tung KY, Giridhar R, Wu WT (2006) Application of binary immobilized Candida rugosa lipase for hydrolysis of soybean oil. J Mol Catal B Enzym 42:32–38

Tomasevica AV, Siler-Marinkovic SS (2003) Methanolysis of used frying oil. Fuel Process Technol 81:1–6

Torres CF, Toré AM, Fornari T, Javier Señoráns F, Reglero G (2007) Ethanolysis of a waste material from olive oil distillation catalyzed by three different commercial lipases: a kinetic study. Biochem Eng J 34(2):165–171

Tynek M, Hazuka Z, Pawlowicz R, Dudek M (2001) Changes in the frying medium during deep-frying of food rich in proteins and carbohydrates. J Food Lipids 8:251–261

Waghmare GV, Rathod VK (2016) Ultrasound assisted enzyme catalysed hydrolysis of waste cooking oil under solvent free condition. Ultrason Sonochem 32:60–67

Wang C, Zhou X, Zhang W, Chen Y, Zeng A, Yin F, Li J, Xu R, Liu S (2011) Study on preparing fatty acids by lipase hydrolysis waste oil from restaurants. Power and Energy Engineering Conference (APPEEC), 2011 Asia-Pasific. March 25–28 2011. IEEE, Wuhan, pp 978–980

Zaharudin NA, Rashid R, Esivan SMM, Othman N, Idris A (2016) Review on the potential use of waste cooking palm oil in the production of high oleic palm oil via enzymatic acidolysis. Jurnal Teknologi 78:85–99

Chapter 10
Potential of Kaffir Lime (*Citrus hystrix*) Peel Essential Oil as a Cockroach Repellent

Sharifah Soplah Syed Abdullah and Muhammad Khairul Ilmi Othman

Abstract The presence of cockroaches in homes and buildings is common. They are one of the most important agents in transmission of bacteria, yeast, protozoa, and parasite worm species to human life either mechanically or biologically. In this work, the potential of Kaffir lime peel toward cockroaches is reported. The peel of Kaffir lime was extracted by hydrodistillation to obtain its essential oil. The repellency of the essential oil was evaluated at different concentrations (0%, 25%, 50%, 75%, 100%, v/v). The duration of the observation for 3 and 6 h was conducted to the cockroaches at lab scale. From the result obtained, the essential oil of Kaffir lime peel exhibited complete repellency at concentration of 50% v/v and above. Such results may be considered as novel findings in the course of searching for potent botanical insecticides against the cockroaches. The result of the present study will provide knowledge and information about Kaffir lime peel as an insect repellent.

10.1 Introduction

Kaffir lime (*Citrus hystrix*) is a spice that has been used for a long time in the Asian countries such as Laos, Indonesia, Malaysia, Vietnam, and Thailand. The valued parts of Kaffir lime are the leaves whereby it has been used to add a distinctive aroma and flavor to food (Fig. 10.1). The fruit is not as extensively used as the leaves. However, the juice of the fruit is good for healthy gums. It has bleaching properties and when mixed with detergent can fight the toughest stains. The rind of the fruit is used in many digestive tonics and blood purifiers. To date, not many studies were reported on the utilization of Kaffir lime peel. It is known that essential oil can be extracted from Kaffir lime leaves. However, the oil yield obtained from the peel is higher than leaves. The essential oils have been used as flavor and fragrance agents, as well as in perfumery (Thavara et al. 2007). Currently, citrus fruits are

S.S.S. Abdullah (✉) • M.K.I. Othman
Section on Bioengineering Technology, Malaysian Institute of Chemical and Bioengineering Technology, Universiti Kuala Lumpur, Alor Gajah, Melaka, Malaysia
e-mail: sharifahsoplah@unikl.edu.my

Z.A. Zakaria (ed.), *Sustainable Technologies for the Management of Agricultural Wastes*, Applied Environmental Science and Engineering for a Sustainable Future, https://doi.org/10.1007/978-981-10-5062-6_10

Fig. 10.1 Kaffir lime (**a**) fruit (**b**) peels and (**c**) leaves

marketed fresh or as purified juice and canned segments, while fruit peel is made in huge quantities and commonly regarded mainly as waste. For this reason, researchers have focused on the utilization of citrus products and by-products (Haroen et al. 2013). One of the potential products is insect repellent which targets the cockroach. Cockroaches have the potential to mechanically carry and transmit many pathogens, such as bacteria, viruses, fungi, and many more. They also serve as potential carriers for bacterial diarrhea and nosocomial infections in hospitals (Cochran 1982). Currently, research regarding cockroach repellents, especially those derived from plant extracts, are quite limited.

10.2 Beneficial Effects of Kaffir Lime Peel

10.2.1 Insect Repellant

Kaffir limes are often used as insect repellant properties in the countries where Kaffir lime was traditionally grown. The citronellol and limonene found in Kaffir limes are very unappealing to most insects (Buatone and Indrapichate 2011).

10.2.2 Health Benefits

Citrus peels and their extracts have been reported to have potent pharmacological activities and health benefits due to the abundance of flavonoids in citrus fruits. Previous study showed that the leaf of this plant contains alkaloid, flavonoid, terpenoid, tannin, and saponin compounds. On the other hand, Kaffir lime leaves and fruit extract were revealed to have antioxidant activity, free radical scavenging ability, antimicrobial activity, and anti-inflammatory activity (Lertsatitthanakorn et al. 2006). With regard to cancer research, Kaffir lime essential oil has been shown to have antiproliferative activity on human mouth epidermal carcinoma (KB) and murine leukemia (P388) cell lines (Manosroi et al. 2006). Furthermore, extract of

Kaffir lime leaf showed cytotoxic effect against HL60 (promyelocytic leukemia), K562 (chronic myelocytic leukemia), Molt4 (lymphoblastic leukemia), and U937 (monocytic leukemia) cells (Chueahongthong et al. 2011). Tunjung et al. (2015) reveal that Kaffir lime extract reduces the viability of cervical and neuroblastoma cell lines and may have potential as anticancer compounds.

Oral Health: Bad breath is a health problem occasionally encountered among adults and children. A readily dissolved edible herbal film containing lyophilized powder of the 80% ethanolic extract of guava leaves (*Psidium guajava* Linn.), Kaffir lime oil (*Citrus hystrix* DC.), and other ingredients was developed (Srisukh et al. 2006). Traditionally, the leaves can be directly rubbed onto the gums to promote good oral health and eliminate harmful bacteria that can build up in the mouth.

Detoxify the Blood The oil of Kaffir limes is often mixed in various decoctions for those suffering from blood-borne illnesses or chronic blood-related diseases. The unique mix of volatile compounds is known to eliminate those pathogens or foreign agents in the blood while also helping the liver and lymphatic system strain out dangerous substances and improve your overall health (www.organicfacts.net/health-benefits/fruit/Kaffir-lime.html).

Digestive Issues There are a number of components found within Kaffir limes that are also found in lemongrass and related herbs. These organic constituents are anti-inflammatory in nature, but they are also stimulating for the digestive system. If you are suffering from constipation or indigestion, some Kaffir lime decoction can clear your symptoms up and get your bowels back on a regular track. This can help prevent more serious gastrointestinal issues in the future, such as colorectal cancer, hemorrhoids, or gastric ulcers (www.organicfacts.net/health-benefits/fruit/Kaffir-lime.html).

Lower Inflammation Kaffir limes can be a very effective remedy for those suffering from rheumatism, arthritis, edema, gout, or some other inflammatory condition (Luangnarumitchai et al. 2007). The juice, leaves, or oil extracts topically can be applied on the area where you are experiencing discomfort or pain. It is also recommended to consume the fruit and the juices to enjoy a similar effect. This anti-inflammatory effect also makes Kaffir lime juice beneficial for headaches and migraines.

Stress Reduction Although most people do not think of Kaffir limes as being particularly useful in aromatherapy, the oil extracted from these powerful fruits can be used aromatically with great effect. Inhaling these soothing vapors can calm the body and mind if you suffer from anxiety or various nervous disorders (Hongratanaworakit and Buchbauer 2007).

Immune System The antibacterial and antioxidant qualities of Kaffir lime make them powerful tools to boost the immune system. Not only does the topical application prevent infections and bacteria from accumulating on the skin, but when consumed, Kaffir limes can help prevent a wide variety of gastrointestinal illnesses and stimulate the immune system via antioxidant effects (Raksakantong et al. 2012).

10.2.3 Cosmetics

Kaffir lime juice and extracts are mixed into many cosmetic and bath products for its wonderful smell, as well as its antioxidant properties. Some of the acids found in Kaffir limes can help to neutralize free radicals, the dangerous by-products of cellular respiration that can cause cell mutation or apoptosis, as well as cancer. Antioxidant compounds also slow the breakdown of cells and minimize the appearance of age marks, scars, and pimples (Lertsatitthanakorn et al. 2006). One of the less well-known applications of Kaffir lime juice and leaves is in the hair. You can apply decoctions and mixtures to the scalp and hair to slow the onset of male pattern baldness and strengthen the follicles of the hair. This also moisturizes the skin to prevent dandruff and improves the appearance and shine (www.organicfacts.net/health-benefits/fruit/Kaffir-lime.html).

The essential oil in the leaves is extracted and used for various purposes. It is used in many bath products such as soaps and shampoos. The oil is a great hair and scalp cleanser. The aroma of the leaves is rejuvenating. It is believed to have a positive effect on the mind and the body and leaves one with positive thoughts. The oil is also infused in deodorants and body sprays for that extra zing. The oil is also used in tonics which aid in digestion and purify the blood.

10.3 Extraction Methods

The common methods of extraction for essential oils are hydrodistillation and steam distillation. As the most common traditional extraction methods are less effective and energy intensive, alternative methods such as microwave-assisted hydrodistillation (MAHD) and microwave steam distillation (MSD) were developed and utilized.

10.3.1 Hydrodistillation

Hydrodistillation is one of the most traditional methods used for extraction. In this method, the raw material is completely sunk in water, whereby it is boiled by applying direct heat in closed steam jacket and closed steam coil or open steam coil (Fig. 10.2). The primary characteristic of this process is the direct contact between boiling water and plant raw material. When the still is heated by direct fire, adequate precautions are needed to prohibit the charge from overheating. When a steam jacket or closed steam coil is used, there is less danger of overheating. On the other hand, when open steam coils is used, this danger is prevented. However, with an open steam, precaution steps must be considered to prevent accumulation of condensed water within the still. Therefore, the still should be well insulated. The plant

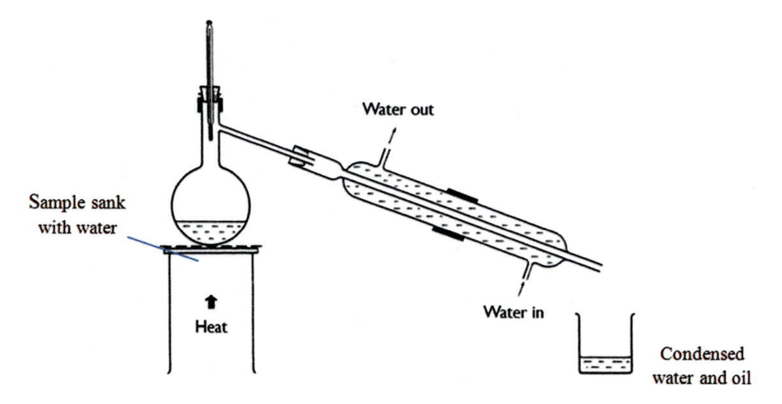

Fig. 10.2 Example of hydrodistillation setup in laboratory scale

material in the still must be movable as the water boils; otherwise load of dense material will sink to the bottom and become thermally degraded (Kidane 2016).

10.3.2 Steam Distillation

The principle of steam distillation is similar with hydrodistillation where the material is being heated so the part of material constituent independently exerts its own vapor as a function of temperature as if the other constituent were not present. But unlike the hydrodistillation, the steam distillation does not require the material to be directly immersed in water. The primary characteristic of this extraction process is that materials and water are not directly in contact with one another. When the steam is passed through the organic material, tiny pockets holding the essential oils will be forced to release the essential oil molecules without doing any damage to these delicate components. The distillate obtained will contain a mixture of water vapor and essential oils which finally return to their liquid form in the condensing apparatus.

10.3.3 Microwave-Assisted Hydrodistillation (MAHD)

Microwave-assisted distillation (MAHD) is the process of heating water in contact with sample using microwave energy to partition the compound from the sample constituent into the vapor. The MAHD process is based on conventional hydrodistillation system with the exception that microwave energy is used during the heating process (Fig. 10.3). The MAHD method was proven to offer important advantages over normal hydrodistillation such as shorter extraction time, better yields, and environmental impact. The energy cost is appreciably higher for performing

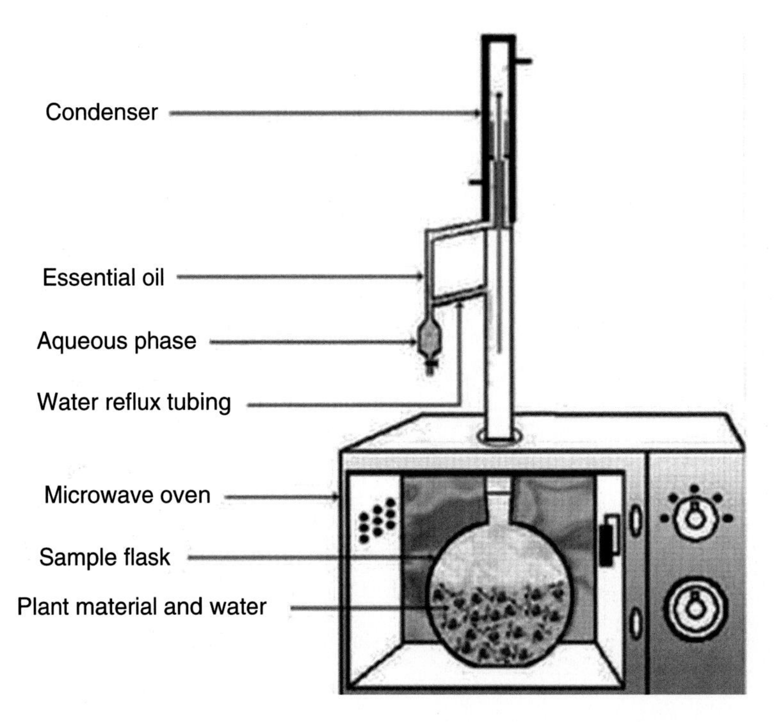

Fig. 10.3 Microwave-assisted hydrodistillation (MAHD) (Jeyaratnam et al. 2016)

hydrodistillation than that required for rapid MAHD extraction. The heating process is based on the molecular motions of the polar molecules and ions inside the solvent and vegetal matrix. It is strongly influenced by the dielectric constants of the solid-liquid-vapor system, developed by process evolution. This heating way realizes a more homogeneous temperature distribution at plant powder suspension level (Koşar et al. 2007). Compared to the hydrodistillation process, pressure difference occurring between the inner and outer side of the plant cells results in a higher effective mass transport coefficient. The microwave energy was absorbed by water and sample that help vaporization process of essential oil. Practically, improvement of the extraction process is a result from the breakdown of external cell wall (Chemat et al. 2005).

10.3.4 Microwave Steam Distillation (MSD)

In this new approach of the improved microwave steam distillation (MSD), the important aspect of the extraction reactor is that only the material that wants to be extracted is submitted to the microwave irradiation, resulting in "hot spots" by selective heating (Fig. 10.4). Because essential oil present inside the sample has a significantly higher dielectric loss than the surrounding steam, steams which do not absorb microwaves flow through the lavender flowers which directly absorb

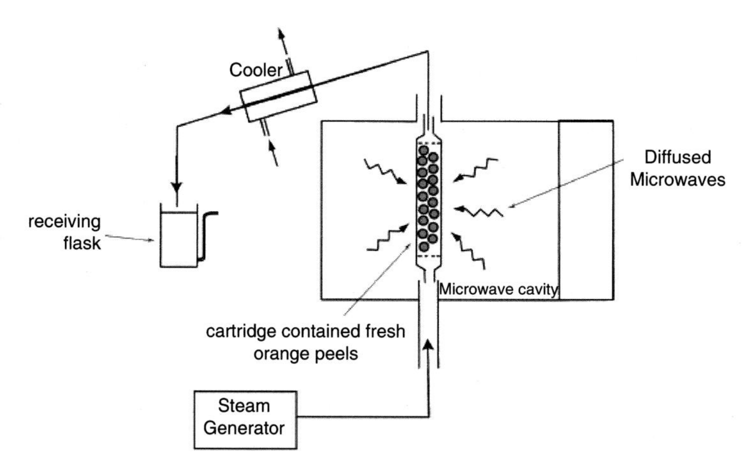

Fig. 10.4 Microwave steam distillation (MSD) (Sahraoui et al. 2008)

microwaves. It is already known that only water in a liquid state absorbs microwaves, but steam and ice do not absorb microwaves because in the gas state the molecules are too far from each other to have frictions and in the solid state the molecules are not free to move and rotate to heat (Metaxas and Meredith (1993)). However, the extraction is conducted in the area of the sample which is continuously heated by microwaves, resulting in a higher local temperature and hence increased extraction rates. This is different compared to the MAHD setup where all the microwave energy is absorbed by water to heat and vaporize and only a fraction is absorbed by the essential oils inside the sample.

10.4 Case Study: Cockroach Repellency Activity of Essential Oil from Kaffir Lime Peel

The cockroaches used were *Dubia roaches* species which were purchased from a local shop in Kuala Lumpur, Malaysia. A required number of adult cockroaches were collected and kept in a plastic container, which was perforated for aeration. These insects were starved for 1 day for experiment. For the present study, an apparatus was improvised outside the laboratory with the following compartments: middle plastic chamber from which projects two PVC tubes (15 cm each). The end of each plastic tube was fitted to two containers and marked as A and B, respectively (Fig. 10.5). Ten starved cockroaches were introduced into the middle plastic chamber. Container A was taken as the treated side while container B as the control side. Biscuit powder that acted as food for the cockroaches was coated with 1 mL of different concentrations of the essential oil extracts, i.e., 0–100% (v/v) oil/ethanol, and placed in container A, while biscuit powder in container B was not treated (Thavara

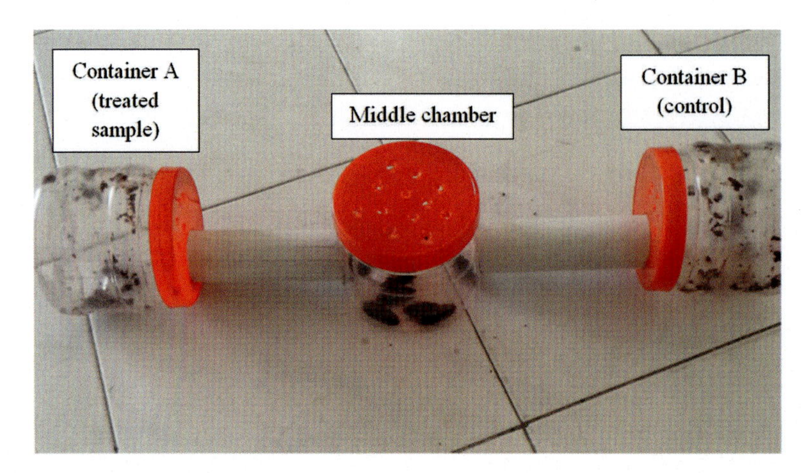

Fig. 10.5 Experimental setup for the repellent test

et al. 2007). The cockroaches located in the treated and control areas were carefully observed and counted at 3 and 6 h after treatment. The extent of repellency of essential oil extracted from the Kaffir lime peel was calculated using the following relationship (Sakuma and Fukami 1985):

$$\text{Yield}(\%) = \frac{\text{Volume of oil}(\text{ml}) \times \text{Specific gravity of oil}\left(\frac{\text{g}}{\text{ml}}\right)}{\text{Weight of sample}(\text{g})} \times 100$$

$$\text{EPI} = \frac{NT - NC}{NT + NC} \text{PC} = \left[1 - \frac{NT}{NT + NC}\right] \times 100$$

EPI: Excess proportion index
NT: Number of insects trapped in the chemical-treated test chamber
NC: Number of insects trapped in the control test chamber
PC: Percentage repellency (percentage of cockroach trapped in control test chamber)

From the evaluation, the percentage of the repellency increased as the oil concentration increased whereby the test cockroaches exhibited 100% repellency against Kaffir lime peel oil at concentration of 50% and above (Fig. 10.6). At 50% essential oil concentration, increased repellency properties were observed when the contact time was prolonged from 3 h (94.4% repellency) to 6 h (100% repellency). This can be attributed to the high concentration of Kaffir lime peel oil that stimulates the cockroaches to leave the treated area. No repellency activity was observed in container B that acted as control (even after 6 h of exposure).

On the other hand, excessive proportion index (EPI) value gives an idea about the repellency or attractancy of chemical substance against an animal tested. The plant extract showed both attractant and repellent activity. The repellency index was classified as EPI <1 which indicates repellency; if EPI = 1 indicates neutral, and if EPI

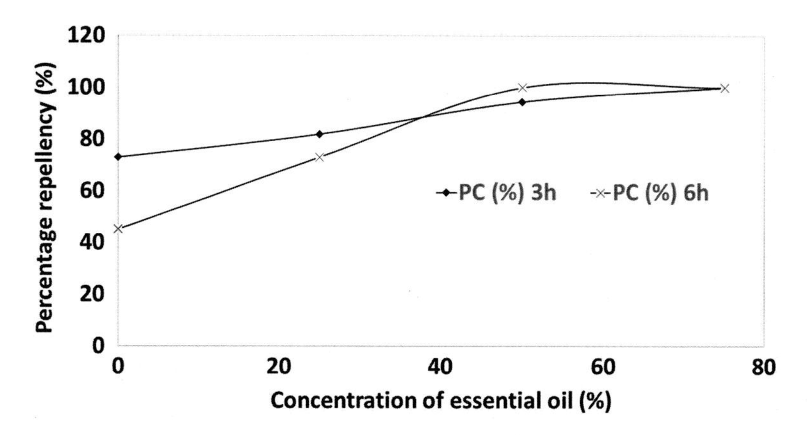

Fig. 10.6 Percentage repellency of cockroaches toward Kaffir lime peel oil

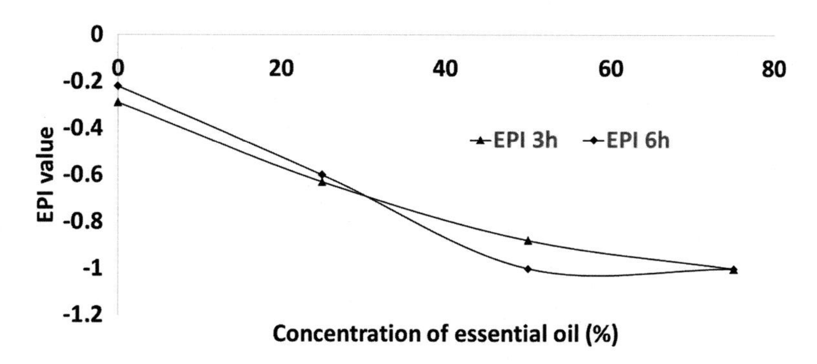

Fig. 10.7 EPI of Kaffir lime peel oil toward the cockroaches

>1, the material used is attractant in nature. It is apparent from the data in Fig. 10.7 that all the Kaffir lime peel oil concentration showed considerable repellent property. The negative value of EPI indicates that number of cockroaches trapped in the essential oil-treated container is less than the number of cockroaches trapped inside container B (control). This implies that EPI and percentage repellency have inverse relationship as clearly indicated in Figs. 10.6 and 10.7. In this experiment, as the concentration of Kaffir lime peel oil increases, the EPI value gradually decreases which thereby gives a gradual hike in the values of percentage repellency. The major reason of Kaffir lime peel oil that has the repellent effect on the cockroach is due to the main composition of the oil itself. The Kaffir lime peel oil has major citronellal component composition (Nerio et al. 2010). The citronellal is also the main component in plant-based insect repellent. Some study shows that sometimes the reason for these types of insect behavior might be due to the presence of the sense of smell stimulus. The presence of volatile compounds having strong odor could block the tracheal respiration of the cockroach leading to their death.

10.5 Conclusion

In conclusion, the essential oil derived from Kaffir lime exhibited complete repellency (100%) at 50% concentration and above. Such results may be considered as novel findings in the course of searching for potent botanical insecticides against the cockroaches. The result of the present study would provide knowledge and information about Kaffir lime peel as an insect repellent.

References

Buatone S, Indrapichate K (2011) Protective effects of mintweed, kitchen mint and Kaffir lime leaf extracts against rice weevils, Stitophilus oryzae L., in stored milled rice. Bioinfo Publ Int J Agric Sci 3(3):975–3710. Available at: http://www.bioinfo.in/contents.php?id=26. Accessed 18 Dec 2016

Chemat S et al (2005) Microwave-assisted extraction kinetics of terpenes from caraway seeds. Chem Eng Process Process Intensif 44(12):1320–1326

Chueahongthong F et al (2011) Cytotoxic effects of crude kaffir lime (Citrus hystrix, DC.) leaf fractional extracts on leukemic cell lines. J Med Plants Res 5(14):3097–3105. Available at: http://www.academicjournals.org/JMPR. Accessed 20 Jan 2017

Cochran DG (1982) "Cockroaches," technical report. World Health Organization, Geneva

Haroen U, Marlida Y, Budianyah A (2013) Extraction and isolation phytochemical and antimicrobial activity of Limonoid Compounds from Orange waste juice. Pak J Nutr 12(8):730–735

Hongratanaworakit T, Buchbauer G (2007) Chemical composition and stimulating effect of Citrus hystrix oil on humans. Flavour Fragr J 22(5):443–449. Available at: http://doi.wiley.com/10.1002/ffj.1820. Accessed 17 Dec 2016

Jeyaratnam N, Nour AH, Akindoyo JO (2016) The potential of microwave assisted hydrodistillation in extraction of essential oil from Cinnamomum Cassia (cinnamon). ARPN J Eng Appl Sci 11(4):2179–2183

Kidane M (2016) Extraction and characterization of essential oil from Eucalyptus leaves using steam distillation. Addis Ababa University

Koşar M et al (2007) Comparison of microwave-assisted hydrodistillation and hydrodistillation methods for the fruit essential oils of Foeniculum Vulgare. J Essent Oil Res 19(5):426–429

Lertsatitthanakorn P et al (2006) In vitro bioactivities of essential oils used for acne control. Int J Aromather 16(1):43–49

Luangnarumitchai S, Lamlertthon S, Tiyaboonchai W (2007) Antimicrobial activity of essential oils against five strains of Propionibacterium acnes. Mahidol Univ J Pharm Sci 34(4):60–64

Manosroi J, Dhumtanom P, Manosroi A (2006) Anti-proliferative activity of essential oil extracted from Thai medicinal plants on KB and P388 cell lines. Cancer Lett 235(1):114–120

Metaxas AC, Meredith RJ (1993) Industrial microwave heating. Peregrinus Ltd., London

Nerio LS, Olivero-Verbel J, Stashenko E (2010) Repellent activity of essential oils: a review. Bioresour Technol 101(1):372–378. Available at: http://www.sciencedirect.com/science/article/pii/S0960852409009468. Accessed 23 Oct 2015

Raksakantong P, Siriamornpun S, Meeso N (2012) Effect of drying methods on volatile compounds, fatty acids and antioxidant property of Thai kaffir lime (Citrus Hystrix D.C.) Int J Food Sci Technol 47(3):603–612. Available at: http://doi.wiley.com/10.1111/j.1365-2621.2011.02883.x. Accessed 17 Dec 2016

Sahraoui N et al (2008) Improved microwave steam distillation apparatus for isolation of essential oils. Comparison with conventional steam distillation. J Chromatogr A 1210(2):229–233

Sakuma M, Fukami H (1985) The linear track olfactometer: an assay device for taxes of the german cockroach, blattella germanica (linn.) toward their aggregation pheromone. Appl Entomol Zool 20:387–402

Srisukh V et al (2006) Readily-dissolved edible herbal film for suppression of bad breath. Thai J Phytopharm 13(1)

Thavara U, Tawatsin A, Payu Bhakdeenuan PW, Boonruad T, Bansiddhi J, Pranee Chavalittumrong NK, Siriyasatien P, Mulla MS (2007) Repellent activity of essential oils against cockroaches (dictyoptera: blattidae, blattellidae and blaberidae) in Thailand. Southeast Asian J Trop Med Publ Health 38(4):663–673. www.organicfacts.net/health-benefits/fruit/Kaffir-lime.html. Accessed on 1 Dec 2016

Tunjung WAS et al (2015) Anti-cancer effect of kaffir lime (Citrus Hystrix DC) leaf extract in cervical cancer and neuroblastoma cell lines. Procedia Chem 14:465–468

Chapter 11
Pineapple Waste Utilization as a Sustainable Means of Waste Management

Zainab Rabiu, Fatima U. Maigari, Umma Lawan, and Zulaihatu Gidado Mukhtar

Abstract Pineapple waste contains various substances that are valuable for the development of new and emerging technologies, nutraceuticals, food, pharmaceuticals, as well as biogas and bioethanol production. Bromelain extraction from pineapple waste is a very highly looked into area, while dietary fibers and phenolic antioxidants could be used as impending nutraceutical resource, capable of offering significant low-cost nutritional dietary supplement for low-income communities. The booming market of functional food has created a vast vista for utilization of natural resources. In this regard, cheap substrates, such as pineapple wastes, have promising prospect. Thus, environmentally polluting by-products could be converted into products with a higher economic value than the main product; hence with sustainable utilization of pineapple waste and with application of novel scientific and technological methods, valuable products from pineapple wastes could be obtained.

Z. Rabiu (✉)
Institute of Bioproduct Development, Universiti Teknologi Malaysia, Johor Bahru, Johor, Malaysia

Department of Biochemistry, Yusuf Maitama Sule University, Kano, Kano State, Nigeria
e-mail: lalabrabiu@yahoo.com

F.U. Maigari
Department of Biochemistry, Gombe State University, Gombe, Nigeria

U. Lawan
Department of Biochemistry, Yusuf Maitama Sule University, Kano, Kano State, Nigeria

Z.G. Mukhtar
Department of Science Laboratory Technology, School of Technology, Kano State Polytechnic, Kano, Kano State, Nigeria

Z.A. Zakaria (ed.), *Sustainable Technologies for the Management of Agricultural Wastes*, Applied Environmental Science and Engineering for a Sustainable Future, https://doi.org/10.1007/978-981-10-5062-6_11

11.1 Introduction

Pineapple (*Ananas comosus L. Merr*) is an important tropical and subtropical plant widely cultivated in many places including the Philippines, Thailand, Malaysia, Mexico, South Africa, Costa Rica, Nigeria, Brazil, and China. It is the leading edible member of the family *Bromeliaceae*, where its fruit juice is the third most preferred worldwide after orange and apple juices (Cabrera et al. 2001). Besides agricultural uses such as nutritional food, this plant is also known for its wide array of pharmacological properties such as antibacterial (Kataki 2010), antihyperlipidemic (Xie et al. 2005), anti-dysuria (Sripanidkulchai et al. 2001), antidiabetic (Xie et al. 2006) and antitumor (Song 1999). Kalpana, Sriram Prasath and Subramanian (2014) reported on the evaluation of antidiabetic and antioxidant properties of *Ananas comosus*' leaves on streptozotocin (STZ)-induced experimental rats.

11.2 Pineapple Waste

Pineapple waste consists of residual skin, peel, pulps, stem, and leaves which are by-products of the pineapple processing industries (Fig. 11.1). It is mostly generated from poor handling of fresh fruit, storage, or lack of good and reliable transportation system (Praveena and Estherlydia 2014). Improper management of these

Fig. 11.1 (**a**) Drií pineapple crown, (**b**) dried pineapple leaves, (**c**) dried pineapple stalk, (**d**) pineapple biomass collection from the plantation

wastes would result in the deterioration of environmental quality which can be attributed mainly to the degradation of the sugar-rich contents.

11.2.1 Waste Management

Waste management or waste disposal includes all measures involved in managing unwanted or discarded materials from inception to final stage of disposal. Management of solid waste is the second most disturbing issue affecting developing countries after water quality (Senkoro 2003). Pineapple waste is an agricultural waste which is generally described as waste produced from farming activities; it can be from natural sources (organic) or un-natural sources (inorganic). Major industrial activities such as food and agricultural-based account for about 30% of total industrial waste generated including liquid, residues, and refuse (Ashworth and Azevedo 2009).

11.2.2 Waste Management Techniques

Techniques employed in managing waste include the following methods: landfilling, incineration, pyrolysis and gasification, composting, and anaerobic digestion. Pineapple waste can be sustainably managed using the thermochemical or the biological conversion method. This includes pyrolysis and gasification as well as fermentation. In developing countries like Nigeria, pyrolysis is used in producing charcoal for domestic cooking and as a waste management technique in pineapple management to produce char, gases, and bio-oil, which can be utilized in the production of other high-value-added products. Through gasification, pineapple waste can also be managed in a sustainable way, to produce syngas and to produce combustion fuels, electricity, and other renewable technologies. Anaerobic digestion can be used in waste management of pineapple under controlled conditions.

11.3 Various Applications of Pineapple Waste

11.3.1 Energy Generation and as Carbon Source

The potential of using pineapple waste as feedstock for energy generation has been looked into for quite some time. Oranusi et al. (2015) reported the generation of biogas (71% CH_4, 18% CO_2, 7.0% N_2, 1.5% H_2, 1.5% H_2S, 1% O_2) using co-digestion of pineapple peels with food waste (1:1) where cattle rumen was used as the inoculums. It was also reported that the yield of biogas is higher and the activity is faster in pineapple biomass compared to watermelon biomass. This can be attributed to the high fermentable sugar content in pineapple biomass, resulting in swift action by hydrolytic bacteria, relative to the watermelon biomass with more fibrous tissues (Mbuligwe and Kassenga 2004). Bio-methanation of fruit wastes is one of the most suitable methods for waste treatment as it both adds energy in the form of methane and also results in a highly stabilized effluent with almost neutral pH and odorless property. Bardiya et al. (1996) reported the semicontinuous anaerobic digestion of pineapple waste that resulted in up to 1682 ml/day of biogas with maximum methane content of 51%. Use of different pineapple peel amounts also resulted in biogas yields ranging from 0.41 to 0.67 m^3/kg volatile solids with methane content of 41–65% (Rani and Nand 2004). Volatile fatty acids (acetic, propionic, butyric, i-butyric, valeric) and methane have been produced from solid pineapple waste, at higher alkalinity. Up to 53 g of volatile fatty acids were produced per kg of pineapple waste (Babel et al. 2004). Pineapple waste has been utilized as a carbon substrate to produce hydrogen gas from municipal sewage sludge (Wang et al. 2006) as well as the production of cellulose by *Acetobacter xylinum* (Kurosumi et al. 2009). In a research conducted by Vijayaraghavan et al. (2007) also determined that the use of 15% pineapple peel in the mixed fruit peel waste resulted in the generation of bio-hydrogen gas at 0.73 m^3/kg of volatile solid destroyed.

11.3.2 Antioxidant Activity

Pineapple waste has been found to have a high content of phytochemical activity and antioxidant capacity that can be harnessed and utilized in various ways. Oliveira et al. (2009) reported that the use of methanolic extract from pineapple waste containing total phenolic contents of 10 mg/g GAE showed substantial nutritional, therapeutical capability and antioxidant activities (DPPH free radical scavenging and superoxide anion scavenging properties). Phenolic compounds such as myricetin, salicylic acid, tannic acid, *trans*-cinnamic acid, and *p*-coumaric acid were determined in the high dietary fiber powder based from pineapple shell (Larrauri et al. 1997). The FRAP value for pineapple peel has been reported as 2.01 mmol/100 g wet weight (Guo et al. 2003). It was also proposed that phenolic antioxidant properties from pineapple waste may be converted to more potent compounds by cytochrome P4502C9 isozyme in vitro (Upadhyay et al. 2009). Other researchers such as Tawata and Upadhyay (2010), Chompoo et al. (2011), and Upadhyay et al. (2011) evaluated the anti-inflammatory and antidiabetic potential of pineapple stem waste. Ongoing research on phytochemicals from pineapple peel and leaf also showed a high antioxidant activity with high phenolic compounds. The leaf also has significant amount of phytosterol content, particularly beta-sitosterol, stigmasterol, and campesterol. Furthermore, the highest amount of phenolic from pineapple peel was extracted in 30 min using 75% ethanol at 75 °C (unpublished data). Different extraction methods also resulted in different yields of total phenolic contents. Sun et al. (2002) carried out the extraction process using 80% acetone followed by base digestion and ethyl acetate extraction, while Gardener et al. (2000) centrifuged the juice before estimating the total phenolic content. Extraction of crude polyphenols using aqueous methanol/ethanol or acetone is quite popular and frequently used

(Gorinstein et al. 2002; Larrauri et al. 1997). The concentrations of solvent used also have impact on the amount of phenolic extracted. Extraction with 50% acetone and 70% ethanol has proven them to be the best solvents for phenolic compounds (Alothman et al. 2009). However, in some cases, concentration of high polar compound is achieved by extracting with hexane before carrying out ethyl acetate extraction. Mathew et al. (2015) reported that dichloromethane extract containing phenolic compounds from pineapple stem and leaves could serve as an alternative eco-friendly source of natural antioxidants.

11.3.3 Pharmaceutical and Food Industry

Bromelain is a proteolytic enzyme present in the stem of pineapple, known as stem bromelain, and also in fruit. It is widely used in pharmaceutical and food industries mostly as a tenderizer and a dietary supplement (Hebbar et al. 2008). It is a crude extract of pineapple that contains, among other components, various closely related proteinases, demonstrating, in vitro and in vivo, anti-edematous, anti-inflammatory, antithrombotic (Bhui et al. 2009), and fibrinolytic activities, and has potential as an anticancer agent (Chobotava et al. 2009). Studies on bromelain extraction and purification from crude aqueous extract of pineapple (core, peel, crown, and the extended stem) use modernistic approaches to produce potential phyto-medical compounds (Hebbar et al. 2008); (Manzoor et al. 2016). Bromelain, unlike papain, does not disappear as the fruit ripens. Crude commercial bromelain from pineapple stem has been purified by successive use of ion-exchange chromatography, gel filtration, and ammonium sulfate fractionation (Murachi et al. 1964). Purifications of bromelain from crude extract have been reported to proceed via two-phase aqueous system (Babu et al. 2008), metal affinity membranes (Nie et al. 2008), and also two-phase partitioning and collagen hydrolysis from pineapple peel (Ketnawa et al. 2010), Bromelain has shown a distinct pharmacological promise as an inflammatory agent, as a platelet aggregation inhibitor, as having fibrinolytic activity and skin debridement properties, and as an antitumor agent.

11.3.4 Citric Acid and Lactic Acid Production

Pineapple waste has been utilized in food industries to enhance taste (citrus). Kumar et al. (2003) reported the use of pineapple waste as a substrate to produce citric acid by *A. niger* under solid state fermentation where variation in the methanol concentration resulted in yield increase from 37.8% to 54.2%. Fungal production of citric acid has been investigated using the yeast *Yarrowia lipolytica* in producing citric acid under solid state fermentation conditions using pineapple waste (from local juice manufacturer) as the sole substrate (Imandi et al. 2008). Tran and Mitchell (1995, 1998) highlighted the higher production yield of citric acid (16.1 g/100 g pineapple waste) using *Aspergillus foetidus* ACM 39969 = FRR 3558 compared to citric acid yield obtained when other types of waste such as apple pomace, rice, or wheat brans were used. Lactic acid production from pineapple waste is applied in both food industries (as an acidulant and preservative) and nonfood industries (as a curing agent, mordant, and moisturizer). Idris and Suzana (2006) used liquid pineapple waste as substrate to produce lactic acid by fermentation using *Lactobacillus delbrueckii* under anaerobic conditions for 72 h. Calcium alginate served as the immobilization matrix to produce maximum yield of 0.78–0.82 g lactic acid/g glucose under different conditions of temperature and pH. Production of lactic acid from pineapple waste was also investigated using fungal species such as *Rhizopus oryzae* and *Rhizopus arrhizus* that produced 14.7 g/L and 19.3 g/L, respectively (Jin et al. 2005). Pineapple syrup has also been used in the production of lactic acid, using *Lactobacillus lactis* and enzyme invertase to hydrolyze sucrose into glucose and fructose; this bio-waste acts as a low-cost substrate (Ueno et al. 2003).

11.3.5 Production of Ethanol

Pineapple waste can be further metabolized to produce ethanol. *Saccharomyces cerevisiae* and *Zymomonas mobilis* are some examples of the microorganisms used at industrial scale for this purpose. However, due to the low availability of fermentable sugars, pretreatment step using enzymes (cellulase, hemicellulase) is necessary. Both organisms were capable of producing about 8% ethanol from pineapple waste in 48 h after pretreatment with enzymes (Ban-Koffi and Han 1990). However, no pretreatment of the pineapple waste is required when respiration-deficient strain of *Saccharomyces cerevisiae* ATCC 24553 was used for continuous ethanol production from pressed juice of pineapple cannery waste. At a dilution rate of 0.05 h^{-1}, the ethanol production was 92.5% of the theoretical value (Nigam 1999a, b). Immobilization of the yeast in k-carrageenan increased the volumetric ethanol productivity by 11.5 times higher relative to yeast cells at a dilution rate of 1.5 h^{-1} (Nigam 2000). Another study using *Zymomonas mobilis* ATCC 10988 using a mixture of pineapple cannery waste and juice of rotten or discarded fruit as substrate resulted in ethanol production of 59.0 g/L without supplementation and regulation in pH (Tanaka et al. 1999).

11.3.6 Heavy Metal Removal from Waste/Wastewater

Pineapple fruit residues also act as effective biosorbent to remove toxic metals like Hg, Pb, Cd, Cu, Zn, and Ni where the addition of phosphate groups increased the adsorbent capacities at lower pH (Senthilkumaar et al. 2000). Citric acid obtained from the fermentation of pineapple wastes using *A. niger* have also been used to leach heavy metals from contaminated sewage sludge (Dacera and Babel 2008) prior to its delivery to the landfill site (Dacera et al. 2009). Pineapple waste water has also been used as cheap substitute of nutrients for *Acinetobacter haemolyticus*, a locally isolated Cr(VI)-resistant reducing aerobic bacterium for the bioreduction process of Cr(VI) to Cr(III) (Zakaria et al. 2007). Pineapple stem is used from aqueous solution as low-cost adsorbent to remove basic dye (methylene blue) by adsorption (Hameed et al. 2009). Pineapple leaf powder has been used in an aqueous solution as a bio-adsorbent of methylene blue in an unconventional way (Weng et al. 2009). Pineapple biomass (crown, leaves, and stem) has shown to be an effective adsorbent in the removal of dyes in wastewaters (Mahmad et al. 2015). Pineapple peel waste activated carbon (PPWAC) has also been shown to be an effective sorbent for removing methylene blue (cationic dyes) from wastewater by increasing its chelating power using sulfuric acid (Yamuna and Kamaraj 2016). There is a need to find alternative cost-effective treatments that are effective in removing dyes from large volumes of effluents, such as biological or combination systems (Robinson et al. 2001). Pineapple waste provides this biological alternative in a sustainable way.

11.3.7 Production of Fibers

Larrauri et al. (1997) have reported that dietary fiber powder prepared from pineapple waste contains 70.6% total dietary fiber (TDF) which is similar to commercial dietary fibers from apple and citrus fruits. However, pineapple has a better sensory property such as neutral color and flavor. Pineapple leaves have been used to make coarse textiles and threads in some Southeast Asian countries (Tran 2006). Alkaline pulping methods were found to be superior over semichemical-mechanical pulping with yields below 40%. A yield of 2.1 g fiber/100 g pineapple pulp waste has been reported by Sreenath et al. (1996). Furthermore, pineapple leaf fibers were investigated in making fiber-reinforced polymeric composites because of high cellulosic content, abundance, and inexpensiveness (Devi et al. 1997; Luo and Netravali 1999; Arib et al. 2006). Studies on the tensile, flexural, and impact behavior of pineapple leaf fiber-reinforced polyester composites as a function of fiber loading, fiber length, and fiber surface modification revealed that the mechanical properties of the composites are superior to other cellulose-based natural fiber composites.

11.3.8 Vinegar Production

Conversions of pineapple peel to vinegar have been reported by Praveena and Estherlydia (2014) through simultaneous fermentation by *Saccharomyces boulardii* and *Acetobacter* where the resulting vinegar were further shown to have phytochemical property and antioxidant capacity. Comparative analysis of the phytochemical screening and antioxidant capacity of vinegar showed that vinegar produced from pineapple peel had higher antioxidant activity (2077 mg acetate equivalent/100 ml) compared to vinegar produced from other fruit wastes. This product can be produced and marketed in large quantities for its therapeutic effects and environment-friendly property over the synthetically produced chemical vinegar. Production of vinegar from pineapple waste using saccharification was reported by Roda, De Faveri, Dordoni and Lambri (2014), with subsequent use of vinegar in the production of disinfectants, food dressing, and preservatives.

11.3.9 Pyroligneous Acid Production

Products of pyrolysis have various applications. They can be used as a source of fuel, as bio-oil, in the production of other high-value-added products, and as sterilizing agents and deodorisers. They have antimicrobial and antioxidant properties and contain other high-value-added products (Rabiu and Zakaria 2016). Pyroligneous acid (PA) from pineapple biomass showed good preliminary indication for use as a wood preservative, exhibiting both antifungal and antitermite properties, the PA biomass showed insignificant inhibition against both *P. sanguineus* and *C. versicolor*, but inhibited the growth of both *A. niger* and *B. theobromae* for 7 days when applied at 70% (v/v) and 100% (v/v) concentrations. Antitermite properties was based on the 100% mortality of *Coptotermes curvignathus* after incubation for a week. GC-MS analysis of the pyroligneous acid biomass showed presence of phenolic compounds and phenol with ortho-substituents, such as 2, 6-dimethoxyphenol and 2-methoxy-4-methylphenol. These compounds have been reported to play a vital role in termiticidal activity (Yahayu et al. 2012).

11.4 Potential Risks and Environmental Benefits

When discharged to the environment, agricultural wastes can be both beneficial and detrimental. The occurrence of agricultural wastes is not restricted to a particular location; rather these wastes are widely distributed where their effect on natural resources such as surface and groundwaters, soil and crops, as well as human health is subsequently felt. Nevertheless, some of their potential risks and environmental benefits are as follows:

1. Proper pineapple waste management can result in economic growth empowerment, employment creation, and revenue for the community or countries. This however depends much on land availability, plant location, scale and choice of technology, and distribution of economic benefits.
2. Reduced health and environmental risks for the communities and surrounding area if the waste is disposed in a sustainable manner. Energy generation from biomass produces small amounts of carbon emissions in the form of CO_2. Other greenhouse gases such as methane and nitrous oxide are produced in small amounts with emissions as low as 2% or less of total emissions. Wihersaari (2010) suggested greenhouse gas reduction up to a maximum of 98% when around 74% of biomass is used in place of fossil fuels.

3. Solid waste like pineapple biomass can be converted to liquid by pyrolysis. Apart from producing the high-value pyrolysis liquid (pyroligneous acid), this approach can reduce transportation cost for waste disposal, notably if the pyrolysis reactor is within the proximity of pineapple waste generation site.
4. Failure in waste management techniques can be attributed to lack of proper management, high cost of mechanization, and improper handling techniques. Improper design, delivery, and operation of thermochemical conversion process pyrolysis/gasification can lead to the emission of acid and greenhouse gases as well as methane production, odor emission, and heavy metal build-up in the final product.

Attaining sustainability in waste management requires an environment-friendly alternative. Such a technique must be effective, efficient, and less costly than many options. Sustainable management is a colossal task over the globe especially in developing nations due to factors like poverty, population explosion, urbanization, and lack of proper funding by the government (UNEP 2002). Techniques employed in waste management such as incinerator, landfill, composting, pyrolysis, and gasification are efficient (El-Fadel et al. 1997; Taiwo 2011). However, if not properly utilized, they may lead to negative impacts on the environment as well as pose a threat to public health.

11.5 Conclusion

In recent years, with the increasing production of pineapple, there will be massive generation of waste products. These wastes can cause environmental pollution problems, if not utilized properly. Therefore there is need for sustainable utilization of pineapple waste into a value-added product, to reduce the wastage, and also produce a viable product, that can be commercialized as an environment-friendly alternative for carbon source utilization this can aid in reducing pollution of the environment. Properly managed waste will provide sustainability and a clean and readily marketable finished product that is environmentally friendly.

References

Adewale M. Taiwo, (2011) Composting as A Sustainable Waste Management Technique in Developing Countries. Journal of Environmental Science and Technology 4 (2):93-102

Alothman M, Bhat R, Karim AA (2009) Antioxidant capacity and phenolic content of selected tropical fruits from Malaysia, extracted with different solvents. Food Chem 115:785–788

Arib RMN, Sapuan SM, Ahmad MMHM, Paridah MT, Zaman HMDK (2006) Mechanical properties of pineapple leaf fibre reinforced polypropylene composites. Mater Des 27:391–396

Ashworth GS, Azevedo P (2009) Agricultural waste nova science publishers', pp 305–309

Babel S, Fukushi K, Sitanrassamee B (2004) Effect of acid speciation on solid waste liquefaction in an anaerobic acid digester. Water Res 38:2417–2423

Babu BR, Rastogi NK, Raghavarao KSMS (2008) Liquid–liquid extraction of bromelain and polyphenol oxidase using aqueous two-phase system. Chem Eng Process 47:83–89
Ban-Koffi L, Han YW (1990) Alcohol production from pineapple waste. World J Microbiol Biotechnol 6:281–284
Bardiya N, Somayaji D, Khanna S (1996) Biomethanation of banana peel and pineapple waste. Bioresour Technol 58:73–76
Bhui K, Prasad S, George J, Shukla Y (2009) Bromelain inhibits COX-2 expression by blocking the activation of MAPK regulated NF-kappa B against skin tumor-initiating triggering mitochondrial death pathway. Cancer Lett 282(2):167–176
Cabrera HAP, Menezes HC, Oliveira JV, Batista RFS (2001) Evaluation of residual levels of benomyl, methyl parathion, diuron, and vamidothion in pineapple pulp and bagasse (Smooth cayenne). J Agric Food Chem 48:5750–5753
Chobotava K, Vernallis AB, Majid FAA (2009) Bromelain's activity and potential as an anti-cancer agent: current evidence and perspectives. Cancer Lett 290:148–156
Chompoo J, Upadhyay A, Kishimoto W, Makise T, Tawata S (2011) Advanced glycation end products inhibitors from Alpinia zerumbet rhizomes. Food Chem 129(3):709–715
Dacera DDM, Babel S (2008) Removal of heavy metals from contaminated sewage sludge using *Aspergillus niger* fermented raw liquid from pineapple wastes. Bioresour Technol 99:1682–1689
Dacera DDM, Babel S, Parkpian P (2009) Potential for land application of contaminated sewage sludge treated with fermented liquid from pineapple wastes. J Hazard Mater 167:866–872
Devi LU, Bhagawan SS, Thomas S (1997) Mechanical properties of pineapple leaf fiber-reinforced polyester composites. J Appl Polym Sci 64:1739–1748
El-Fadel M, Findikakis AN, Leckie JO (1997) Environmental impacts of solid waste landfilling. J Environ Manag 50:1–25
Gardener PT, White TAC, McPhail DB, Duthie GG (2000) The relative contributions of vitamin C, carotenoids and phenolics to the antioxidant potential of fruit juices. Food Chem 68:471–474
Glossary of Environment Statistics: Series F, No. 67/Department for Economic and Social Information and Policy Analysis, United Nations (1997) UN, New York
Gorinstein S, Martin-Belloso O, Lojek A, Ciz M, Soliva-Fortuny R, Park Y-S, Caspi A, Libman I, Trakhtenberg S (2002) Comparative content of some phytochemicals in Spanish apples, peaches and pears. J Sci Food Agric 82:1166–1170
Guo C, Yang J, Wei J, Li Y, Xu J, Jiang Y (2003) Antioxidant activities of peel, pulp, and seed fractions of common fruits as determined by FRAP assay. Nutr Res 23:1719–1726
Hameed BH, Krishni RR, Sata SA (2009) A novel agricultural waste adsorbent for the removal of cationic dye from aqueous solutions. J Hazard Mater 162:305–311
Hebbar HU, Sumana B, Raghavarao KSMS (2008) Use of reverse micellar systems for the extraction and purification of bromelain from pineapple wastes. Bioresour Technol 99:4896–4902
Idris A, Suzana W (2006) Effect of sodium alginate concentration, bead diameter, initial pH and temperature on lactic acid production from pineapple waste using immobilized *Lactobacillus delbrueckii*. Process Biochem 41:1117–1123
Imandi SB, Bandaru VVR, Somalanka SR, Bandaru SR, Garapati HR (2008) Application of statistical experimental designs of medium constituents for the production of citric acid from pineapple waste. Bioresour Technol 99:4445–4450
Jin B, Yin P, Ma Y, Zhao L (2005) Production of lactic acid and fungal biomass by Rhizopus fungi from food processing waste streams. J Ind Microbiol Biotechnol 32:678–686
Kataki MS (2010) Pharmacol Online 2:308–319
Kalpana MB, Sriram Prasath G, Subramanian S (2014) Studies on antidiabetic activity of *Anans cosmus* leaves in STZ induced diabetic rats. Pharm Lett 6(1):190–198
Ketnawa S, Rawdkuen SF, Chaiwut P (2010) Two phase partitioning and collagen hydrolysis of bromelain from pineapple peel Nang Lae cultivar. Biochem Eng J 52:205–211
Kumar D, Jain VK, Shanker G, Srivastava A (2003) Utilisation of fruits waste for citric acid production by solid state fermentation. Process Biochem 38:1725–1729
Kurosumi A, Sasaki C, Yamashita Y, Nakamura Y (2009) Utilization of various fruit juice as carbon source for production of bacterial cellulose by *Acetobacter xylinum* NBRC 13693. Carbohydr Polym 76:333–335

Larrauri JA, Ruperez P, Calixto FS (1997) Pineapple shell as a source of dietary fiber with associated polyphenols. J Agric Food Chem 45:4028–4031

Luo S, Netravali AN (1999) Mechanical and thermal properties of environment-friendly "green" composites made from pineapple leaf fibers and poly (hydroxybutyrate-co-valerate) resin. Polym Compos 20:367–378

Mahamad MN, Zaini MA, Zakaria ZA (2015) Preparation and characterization of activated carbon from pineapple waste biomass for dye removal. Int Biodeterior Biodegrad 102:274–280

Manzoor Z, Nawaz A, Mukhtar H, Haq I, (2016) Bromelain: methods of extraction, purification and therapeutic applications. Braz Arch Biol Technol 59 (0)

Mathew S, Zakaria ZA, Musa NF (2015) Antioxidant property and chemical profile of pyroligneous acid from pineapple plant waste biomass. Process Biochem 50(11):1985–1992

Mbuligwe SE, Kassenga GR (2004) Feasibility and strategies for anaerobic digestion of solid wastes for energy production in Dares Salaam city, Tanzania. Resour Conserv Recycl 42:183–203

Murachi T, Yasui M, Yasuda Y (1964) Purification and physical characterization of stem bromelain

Nie H, Li S, Zhou Y, Chen T, He Z, Su S, Zhang H, Xue Y, Zhu L (2008) Purification of bromelain using immobilized metal affinity membranes. J Biotechnol 136(S):S402–S459

Nigam JN (1999a) Continuous ethanol production from pineapple cannery waste. J Biotechnol 72:197–202

Nigam JN (1999b) Continuous cultivation of the yeast Candida utilis at different dilution rates on pineapple cannery waste. World J Microbiol Biotechnol 15:115–117

Nigam JN (2000) Continuous ethanol production from pineapple cannery waste using immobilized yeast cells. J Biotechnol 80:189–193

Oliveira AC, Valentim IB, Silva CA, Bechara EJH, Barros MP, Mano CM, Goulart MOFG (2009) Total phenolic content and free radical scavenging activities of methanolic extract powders of tropical fruit residues. Food Chem 115:469–475

Oranusi S, Owolabi JB, Dahunsi SO (2015) Biogas generation from water melon peels, pineapple peels and food wastes. International conference on African development issues CU- ICADI Biotechnology and bioinformatics track

Praveena JR, Estherlydia D (2014) Comparative study of phytochemical screening and antioxidant capacities of vinegar made from peel and fruit of pineapple (*Ananas comosus* l.) Int J Pharma Biol Sci Int J Pharm Biol Sci. 2014 5(4):394–403

Rabiu Z, Zakaria ZA (2016) Pyroligneous acid production from palm kernel shell biomass. J Appl Environ Biol Sci 7(2S):59–62, 2017

Rani DS, Nand K (2004) Ensilage of pineapple processing waste for methane generation. Waste Manag 24:523–528

Robinson T, McMullan G, Marchant R, Nigam P (2001) Remediation of dyes in textile effluent: a critical review on current treatment technologies with a proposed alternative. Bioresour Technol 77(3):247–255

Roda A, De Faveri DM, Dordoni R, Lambri M (2014) Vinegar production from pineapple waste-preliminary –saccharification trials. Chem EngTrans 37:607–612

Senkoro H (2003) Solid waste in Africa: a WHO/AFRO perspective. CWG workshop: solid waste collection that benefits the urban poor. Dar es Salaam

Senthilkumaar S, Bharathi S, Nithyanandhi D, Subburam V (2000) Biosorption of toxic heavy metals from aqueous solutions. Bioresour Technol 75:163–165

Song LL (1999) [Chinese herbs. chapter 8] Shang Hai, Administrant Department of National Chinese Traditional Medicine. pp 296–297 (in Chinese)

Sreenath HK, Sudarshanakrishna KR, Prasad NN, Santhanam K (1996) Characteristics of some fiber incorporated cake preparations and their dietary fiber content. Starch-Starke 48:72–76

Sripanidkulchai B, Wongpanich V, Laupattarakasem P, Suwansaksri J, Jirakulsomchok D (2001) Diuretic effects of selected Thai indigenous medicinal plants in rats. J Ethnopharmacol 75:185–190

Sun J, Chu Y, Wu X, Liu RH (2002) Antioxidant and anti-proliferative activities of common fruits. J Agric Food Chem 50:7449–7454

Taiwo AM (2011) Composting as a sustainable waste management technique in developing countries. J Environ Sci Technol 4(2):93–102

Tanaka K, Hilary ZD, Ishizaki A (1999) Investigation of the utility of pineapple juice and pineapple waste material as low cost substrate for ethanol fermentation by *Zymomonas mobilis*. J Biosci Bioeng 87:642–646

Tawata S, Upadhyay A (2010) Applicability of mimosine as neuraminidase inhibitors. Japan Kokai Tokyo Koho. (Japan Patent)

Tran AV (2006) Chemical analysis and pulping study of pineapple crown leaves. Ind Crop Prod 24:66–74

Tran CT, Mitchell DA (1995) Pineapple waste - a novel substrate for citric acid production by solid state fermentation. Biotechnol Lett 17:1107–1110

Tran CT, Sly LI, Mitchell DA (1998) Selection of a strain of Aspergillus for the production of citric acid from pineapple waste in solid-state fermentation. World J Microbiol Biotechnol 14:399–404

Ueno T, Ozawa Y, Ishikawa M, Nakanishi K, Kimura T (2003) Lactic acid production using two food processing wastes, canned pineapple syrup and grape invertase as substrate and enzyme. Biotechnol Lett 25:573–577

UNEP (2002) International source book on environmentally sound technologies for municipal solid waste management. (IETC) Technical Publication, USA

Upadhyay A, Chompoo J, Kishimoto W, Makise T, Tawata S (2011) HIV-1 integrase and neuraminidase inhibitors from Alpinia zerumbet. J Agric Food Chem 59:2857–2862

Upadhyay A, Uezato Y, Tawata S., Ohkawa H (2009) CYP2C9 catalyzed bioconversion of secondary metabolites of three Okinawan plants. In: Shoun H, Ohkawa H, (eds) Proceedings of 16th international conference on cytochrome P450, Nago, Okinawa, Japan, 2009, pp 31–34

Vijayaraghavan K, Ahmad D, Soning C (2007) Bio-hydrogen generation from mixed fruit peel waste using anaerobic contact filter. Int J Hydrog Energy 32:4754–4760

Wang CH, Lin PJ, Chang JS (2006) Fermentative conversion of sucrose and pineapple waste into hydrogen gas in phosphate-buffered culture seeded with municipal sewage sludge. Process Biochem 41:1353–1358

Weng CH, Lin YT, Tzeng TW (2009) Removal of methylene blue from aqueous solution by adsorption onto pineapple leaf powder. J Hazard Mater 170:417–424

Wihersaari M (2010) Greenhouse gas emissions from final harvest fuel chip Production in Finland. Biomass BioEnergy 28(5):435–443

Xi W, Wang W, Su H, Xing D, Pan Y, Du L (2006) Efffects of ethanolic extracts of Ananas cosmos L. leaves on insulin sensitivity in rats and HepG2. Comp Biochem Physiol C Toxicol Pharmacol 143(4):429–435

Xie WD, Xing DM, Sun H, Wang W, Ding Y, Du LJ (2005) The effects of Ananascomosus L. leaves on diabetic–dyslipidemic rats induced by alloxan and a high-fat/high-cholesterol diet. Am J Chin Med 33:95–105

Yahayu M, Mahmud KN, Mahamad MN, Ngadiran S, Lipeh S, Ujang S, Zakaria ZA (2017) Efficacy of pyroligneous acid from pineapple waste biomass as wood preserving agent. J Teknol 79(4):1–8

Yamuna M, Kamaraj M (2016) Pineapple peel waste activated carbon as an adsorbent for the effective removal of methylene blue dye from aqueous solution. Int J ChemTech Res 9(05):544–550

Zakaria ZA, Zakaria Z, Surif S, Ahmad WA (2007) Biological detoxification of Cr (VI) using wood-husk immobilized Acinetobacter haemolyticus. J Hazard Mater 148:164–171

Printed in the United States
By Bookmasters